THE IMAGINATIONS
OF UNREASONABLE MEN

THE IMAGINATIONS
OF UNREASONABLE MEN

INSPIRATION, VISION, AND PURPOSE
IN THE QUEST TO END MALARIA

BILL SHORE

PUBLICAFFAIRS
New York

Published in the United States by PublicAffairs™,
a member of the Perseus Books Group.

PublicAffairs books are available at special discounts for bulk purchases in the U.S. by corporations, institutions, and other organizations. For more information, please contact the Special Markets Department at the Perseus Books Group, 2300 Chestnut Street, Suite 200, Philadelphia, PA 19103, call (800) 810-4145, ext. 5000, or e-mail special.markets@perseusbooks.com.

Book Design by Trish Wilkinson
Set in 11.5 point Minion Pro

Library of Congress Cataloging-in-Publication Data

Shore, William H.
    The imaginations of unreasonable men : inspiration, vision, and purpose in the quest to end malaria / Bill Shore.
        p.   cm.
    Includes bibliographical references and index.
    ISBN 978-1-58648-764-5 (hardcover)
    1. Malaria—Prevention. I. Title.
RA644.M2S465 2010
614.5'32—dc22   2010023877

First Edition

10 9 8 7 6 5 4 3 2 1

For Rosemary, first, last, and always.

CONTENTS

CONTENTS

INTRODUCTION

A T MY DINING-ROOM TABLE, the glow of two flickering candles illuminates the photograph of a beautiful young woman. In the image she is thirteen years old and sitting attentively at a polished wooden desk. Her skin is coffee brown, her eyes bright and searching, and her dazzling smile and gentle expression hold the promise of a limitless future.

The picture was taken in an Ethiopian village called Yetebon, about a three-hour drive south of Addis Ababa. I was there in 2002 with a delegation of business and philanthropic leaders who support Share Our Strength, the anti-hunger organization my sister and I founded in 1984 following one of Ethiopia's most devastating famines. We started the organization with the belief that everyone has a strength to share in the global fight against hunger and poverty, and that in these shared strengths lie sustainable solutions.

Project Mercy is a U.S.-based nonprofit that seeks to educate and supplement healthy lifestyles for impoverished communities around the world. Its first and main campus is located in Yetebon, where Project Mercy had built schools and kitchens and helped to plant community gardens. In the wake of a famine, the group was in the midst of a new construction project—a hospital. Share Our Strength had gone to Yetebon to partner with Project Mercy and to generate more awareness and resources for its work.

At one point in our trip, a few of us stepped into an English class in the Project Mercy campus school. The teacher asked one child after another to stand and tell us what they wanted to be when they were older. After each child had spoken, and after I had thanked the class for allowing us to visit, one girl said something so quietly that I could hardly hear her. She was the only person who spoke without being called upon. I walked over and knelt down to ask her what she had said. She repeated so that I could hear it: "God bless you."

Like any child, she was shy, but unlike many she did not look away. Something about her presence set her apart. She told me her name. I asked her to write it down for me so that I would know the correct spelling. She searched her notebook for an empty space and carefully formed the letters of the English alphabet she had learned in school, writing "Alima Dari."

We talked for five or ten more minutes. I told her where we were from and why we'd come to visit. I complimented her on her English. She told me about her brothers and about her walk to school, and where her family lived.

Eventually, I rejoined the Share Our Strength group to tour the cattle shed, the gardens and kitchen, and the partially built hospital. When finally it was time for us to leave, all of the children, hundreds of them, lined the road from the school to the main gate. As I walked toward them, I scanned the faces for Alima's. There were close to three hundred children, standing three rows deep. It should have been impossible to find her. In fact I soon realized that it was impossible *not* to find her. She beamed at me, and I waved and yelled "Alima!" I reached across the first row of children and we shook hands again.

On my way back to Addis Ababa, and to the United States, I asked myself why one young woman among so many had made such an impression on me. I didn't fully understand it then, nor do I claim to understand it now. I just knew that it moved me. I was simply delighted to have met Alima. I was struck by the sense that anything was possible for her— or for anyone who was so ready to live life to the fullest. From that day forward I followed her progress. For a little over a year we exchanged letters. I received pictures of her reading her graduation speech. I have had many different experiences in my travels for Share Our Strength, but never have I connected with anyone quite the way I did with Alima.

The following summer, when my colleague Chuck Scofield returned to Ethiopia, I gave him a handwritten letter to deliver to Alima. Though Chuck and I keep in close touch, I didn't hear from him for weeks. Then one morning I received this email:

3

Dear Billy, I have not called because I have been avoiding sharing bad news that I learned with regard to Alima. She died a couple of months ago as a result of TB and cerebral malaria. All at Project Mercy were and are extremely sad about losing such an amazing person. Evidently the hospital in Butajira only treated the TB without realizing that she had the most deadly form of malaria. By the time they got her to the hospital in Addis it was too late. I hate like hell to share this news with you.

I was stunned. I have often been moved by the work Share Our Strength carries out. The organization has frequently worked in difficult circumstances in the aftermath of tragedy and disaster. But this was the first time in nearly two decades that I'd felt a sense of loss that touched me personally. It was the first time I'd experienced the brutal impact of poverty and disease on someone I knew and whom I had come to care for. This one small catastrophe had taken place not in a ravaged landscape but, ironically, in a setting of optimism and hope. With all the new building and progress at Yetebon, it was cruelly incongruous that Alima should have died.

It is tempting to describe Alima's death as senseless, but in a terrible way it makes perfect sense. Nearly 3,000 children die from malaria every day, almost 1 million each year. Malaria is the leading cause of death for children in Africa. Global spending on malaria control at the time of Alima's death was $200 million a year—a drop in the ocean. Perhaps

Alima's death was inevitable. Treatment within twenty-four hours of the onset of malaria symptoms is essential. Unfortunately, lack of sufficient funds had prevented the hospital at Yetebon, a short walk from Alima's classroom, from being finished before she contracted malaria. Although Yetebon now has its hospital, many African towns and villages do not.

Such thoughts swirled in my head in the wake of the news about Alima. They were somewhat despairing. In time, though, I became convinced that Alima's short life was long enough to show that action and inaction each have consequences, that lives hang in the balance when it comes to the generosity and commitment with which we pursue our work. It was long enough to make me aware of the fact that Alima and her classmates were among the most voiceless beings on the planet. Children in their situation around the world are so nearly silent and invisible that there is no economic market for delivering to them the basic goods and services that we take for granted and that they desperately need just to stay alive. Given the huge up-front investment that drug and vaccine development require, there's no profit to be found on a continent where people—potential customers—earn less, on average, than $2 a day. When economic markets fail, the gap is sometimes filled by political markets, or governments responding to a need. When economic and political markets both fail, as they have failed Africa's children, only charity or philanthropy remain as a last resort.

While I was thinking about the lessons that Alima's life signified, I began to wonder who, in the developed world, might be trying to help children like Alima and her class-mates. Was there anyone dedicated, determined, or driven enough to want to try to cure a parasitic disease like malaria, which ravages not New Jersey and California but countries and peoples continents away, who have neither the money to pay for treatment nor the ability even to ask for it? Was there anyone who was possessed by the idea, as I had become, that malaria had to be defeated?

Victory in such a battle does not come easily. Drug and vaccine development requires a huge amount of investment in both time and money, and even those who have both must overcome an incredible number of obstacles.

It was in 2004 that I first began to really think about the teams of researchers working on the malaria vaccine and the specific hurdles facing them. The more I learned about the nature of the disease, the more I realized that conquering it would take more than just time and money, more even than a sense of purpose and persistence. It required a moral vision and imagination: a person or a team stubbornly dedicated to the idea that no child's death should ever be ignored as "inevitable" or "senseless," and abundantly blessed with practical wisdom to tackle a problem that has baffled others.

I asked myself what kind of people, with what qualities of character, would take on a challenge as daunting as climbing Mount Everest was before Tenzing Norgay and Edmund Hillary, or walking on the moon before Buzz Aldrin and Neil Armstrong. Ending malaria may be less visually dramatic than a moon landing, and the beneficiaries—people like Alima's classmates—are less conspicuous. That's exactly why it takes a very unusual person to take on a challenge like this one.

But beyond that, the quest for a malaria cure became emblematic for me of humanitarian endeavors in general, especially those that presented such huge challenges that success had eluded us time and again. The more I discovered about malaria research, the more I became convinced that it held clues to how any quest of this magnitude, with so many attempts and failures behind it, could finally succeed. And the more I looked for answers, the more I came to believe that it is the character of the people doing the work that is the key. Their methods can be baffling and surprising, and sometimes they can even be unreasonable. They are different from the rest of the crowd. And I wanted to understand how they were different.

This book is inspired by Alima's memory. But beyond being a celebration of her, it is a tribute to the quest undertaken by a small number of heroic idealists. It is a tribute to the imaginations of unreasonable men.

CHAPTER I

WHEN GOOD IS
NOT GOOD ENOUGH

Federal officials announced today that scientists had cleared the last major hurdle to development of a vaccine against malaria. . . .

. . . It should now be possible [officials and scientists] said, to mass produce a vaccine that will stimulate immunity against at least one stage of the major form of malaria.

M. Peter McPherson, administrator of the Agency for International Development, expected that a vaccine would be ready for trial in humans within 12 to 18 months and widely available throughout the world within five years.

—Philip M. Boffey, "Malaria Vaccine Is Near,
U.S. Health Officials Say," *New York Times*, August 3, 1984

IN THE FALL OF 2009, I was invited to speak at a gathering of foundation and nonprofit CEOs from Massachusetts. I've spoken to many such groups over the years, usually about nonprofit effectiveness and strategies for reaching scale and sustainability. This was different, a setting unlike any I'd encountered, as was the theme I was asked to address. It became a turning point in my thinking about the ingredients needed to succeed at the kind of work in which we engage.

The invitation came from the Pucker Gallery in Boston, which was showing the work of a renowned potter, the late Brother Thomas Bezanson. His pottery includes tea bowls, vases, and large decorative plates known for their elegant glazes. His work has been collected by the Metropolitan Museum of Art, Boston's Museum of Fine Art, the Smithsonian's Renwick Gallery, and many other prominent institutions.

I know little about pottery and ceramics and had never heard of Brother Thomas. But the gallery owners, Bernie and Sue Pucker, are active in Boston's philanthropic community and we had mutual friends and interests. They asked me to speak on the connections between the spirituality of Brother Thomas's art and spirituality in social justice work. It was virgin territory for me, requiring more than the usual amount of preparation. I visited the gallery several times to learn more about Brother Thomas and his work.

I prepared my talk at the same time I was working on this book. Science and pottery might seem to have little overlap, but what I was finding most exciting in my research for the book were the qualities of character and entrepreneurial strategies that led to discoveries and breakthroughs. They had relevance beyond any single project. They were pertinent to my own life's work of trying to end hunger and to a plethora of seemingly impossible-to-solve issues, such as climate change, education, human rights, and health care. Such qualities and strategies are not always as obvious and familiar as the more concrete external resources we reflexively seek, such as money, technology, expertise, and political support. But that doesn't make them less essential. Like diamonds deep beneath the surface, their scarcity and invisibility make them all the more valuable to capture and bring to light. As Antoine de Saint Exupery's fox said to the Little Prince, "what is essential is invisible to the eye."[1]

Character qualities are especially critical for tackling problems that affect people so politically and economically marginalized that there are no market incentives for solving them. My dozen years on Capitol Hill and in presidential politics, and quarter century in the nonprofit sector, taught that those problems seem never to go away and are the toughest of all to solve: poverty, disease, ignorance, inequality. Traditional approaches always fall short. People who devote their careers to such problems are simultaneously admired and dismissed as idealists. In the absence of a new way of thinking, the frustrating cycle of finding and allocating hard-won resources, whether public or private, toward problems that resources alone can't solve, futilely continues.

Such pathology is discouraging to the increasing number of people who desire to make a difference. They want to give something back, but are not sure how to do so most effectively. The nonprofit sector is growing rapidly and is increasingly diverse. But it seems perennially hobbled by shortages of money and talent and by old traditional ways of doing things. Too often, good intentions stand in for transformational thinking and disciplined strategy. Many well-meaning organizations, efforts, and movements fail to live up to their full potential.

But what is full potential? And how can we discover it? Here, the detour of trying to understand Brother Thomas through his art was profoundly revelatory.

GOOD IS NOT GOOD ENOUGH

To read the basic biographical facts about Brother Thomas Bezanson, you'd think he'd lived an ordinary monastic life. He was born in Nova Scotia in 1929, graduated from the College of Art and Design there at the age of twenty-one, and then earned a degree in commerce from St. Mary's University in Halifax. He'd entered the Benedictine Monastery in Weston, Vermont, by the age of thirty, and earned a doctoral degree in philosophy at Ottawa University in 1968. Brother Thomas became artist-in-residence with the Benedictine Sisters in Erie, Pennsylvania, in 1985, and he died, at the age of seventy-eight, in 2007.[2]

But Brother Thomas was far from ordinary. His work has appeared in dozens of public exhibits in some of the most prominent museums in the world, as sought after today as it was during his lifetime. And then there is this curious fact: Of the first 1,200 pieces he created, Brother Thomas broke 1,100, a ratio he adhered to throughout his life. He routinely destroyed hours' and days' worth of solitary creative effort and disciplined craftsmanship.

Brother Thomas lived by a unique set of standards. Even when his pots were good, they were not good enough, begging the question: Not good enough for whom? Would the flaws even have been noticeable to anyone but himself? Yet, no matter how good they may have been, they were not good enough for Brother Thomas. As much as others may have admired them, they did not represent the version of himself that he was determined to express. They were not true to what he believed to be his highest potential.

It's not that Brother Thomas was aiming for perfection—he was wise enough to know that is unattainable. But he was aiming for the best possible. What distinguished him from his peers, and what accounted for his success, was a more expansive vision of what could be accomplished. Impractical was not a disqualifier. Nor was inconvenient, expensive, or extremely labor-intensive. These were merely obstacles to be worked around or run over. They rendered his quest more difficult, but in no way altered the reality of what he knew to be within reach.

Most important of all, he was not just aiming for the best possible but was holding himself accountable to the highest of standards. In an essay in *Creation Out of Clay*, a collection of his art and writings, Brother Thomas wrote: "If I were a pottery manufacturer, then losing half of my work would be madness. . . . To be unfaithful to my own inner vision of what is beautiful-to-me would be the beginning of an inner lie . . . that would soon render my work inauthentic."[3]

Extensive photos documenting Brother Thomas at work show a man, not surprisingly, as physically strong as he was mentally tough and determined. Large, round, black-rimmed glasses are all that soften the Zeus-like look bestowed by a large and long head, thick tight curls of grey hair, and a speckled full beard. His powerful hands shape the clay into the most delicate-looking vessels, at times so lathered in wet clay, and so sturdy in appearance, that it is hard to tell where his flesh ends and the pot begins.

Working with little else but those hands, and occasionally a stick or knife, Brother Thomas produced a stunning range of art. A fire burned within that was every bit as intense as the fire in his workshop's kiln. And he put it to the same use: hardening his determination to work according to his own vision, no matter what others might have thought.

For Brother Thomas, good was not and never would be good enough. It's an admirable, even inspirational, philosophy. But it might be better suited to the monastery than the marketplace. Breaking 1,100 of every 1,200 pots could also be interpreted as stubborn, eccentric, unrealistic, or unreasonable.

"Good but not good enough" implies a restless and relentless push for more, a refusal to accept what others accept. It borders on hubris that nearly disparages the ease and comfort most of us are content to seek and embrace. But aren't these qualities often embedded somewhere in the foundations of great achievement? Aren't they always?

In the DNA of every great and worthy breakthrough is a gene encoded with the instruction that good is not good enough. It is not only in Brother Thomas's pottery: It was also in Joe DiMaggio's swing of a bat. It is visible in the ceiling of the Sistine Chapel and audible in the strains proceeding from Yo Yo Ma's cello. It was evident in Rosa Parks's belief that a seat in the back of the bus was not good enough, and her refusal to accept one. It prevailed in NASA's determination to reach the moon and Gandhi's determination that India reach independence.

It would be easy to confuse this quality with a classic work ethic, the kind that leads a Boy Scout to an Eagle badge. However, good but not good enough is not just about practicing longer, working harder, or being more competitive. Instead it is a deeply intrinsic drive to achieve what others have dismissed as unachievable, or have simply not been able to even imagine. It's a drive powered by internal vision and compass and indifferent to external expectations, conventional wisdom, skepticism, or even ridicule. It demands a willingness to take risks that often seem unreasonable right up until the moment they succeed.

Eradication of the threat of malaria throughout the world is the kind of challenge that demands unreasonable imagination, a willingness to break a lot of pots before expecting a solution.

The audacious goal of saving the lives of nearly 1 million children a year—the number currently dying from malaria—will require new breakthrough thinking, considering the half century of high but continually shattered hopes in the history of malaria eradication efforts. Brilliant research by dedicated scientists across the globe has taken place over the past fifty years, most of it against the backdrop of the incremental progress that was believed to be all that was possible at the time. And yet the problem persisted, with the number of malaria cases actually rising instead of falling around the world.

What was needed in the seemingly quixotic quest to create and manufacture an effective vaccine for malaria was the stubborn conviction that what could be accomplished was greater than what anyone else in the field had thought possible. Good was not good enough.

Many of those pursuing social change have reached a similar place—a place where incremental progress has led to a frustrating plateau. And then along comes someone who decides to turn the old methods upside-down and do something different. Whether with hunger, health care, housing, schools, or any of dozens of other issues, a dividing line has grown ever brighter: On one side are the many efforts to ameliorate the symptoms of a problem; on the other are extraordinary efforts to attack the root causes and eradicate it altogether. That line marks the difference between those content to stand on good intentions and those willing to risk a public commitment to a specific, often ambitious outcome.

Wealthy donors, foundations, and others are increasingly gathering on one side of that line. It is creating a sea change in the conduct of philanthropy and explains why there is so much emphasis today on more focused investing for impact, strategic management, and technical assistance; measurable outcomes and greater transparency; and the scaling of evidence-based programs. From small family and community foundations to massive institutions like the Kellogg and Ford foundations, there is long overdue reorientation and refocusing underway founded on the impatience that accompanies the idea that good is not good enough.

Share Our Strength went through just such an evolution in thinking and strategy, which gave me a firsthand perspective of what is involved. We had to establish a priority: Was it to feed hungry people, or to address the root causes of what made people hungry and try to eradicate them?

The urgent and immediate needs of people who are hungry often overwhelm the more ambitious target. When you look into the eyes of those who are suffering, whether from sickness, cold, or hunger, or just because they lack opportunities enjoyed by the rest of society, it can sound callous to say, "Sorry, I'm devoting my energy to attacking the root causes of your suffering, but unfortunately can't address your suffering itself."

Share Our Strength began in 1984 as a grant-maker and for many years funded hundreds of anti-hunger organizations across the country and around the world, awarding more than $100 million in grants by 2010. We didn't compete for existing philanthropic dollars but brought new resources into the community. We funded the operating expenses of organizations that no one else wanted to fund.

We were nonbureaucratic, reliable, and loyal to those we funded. We received great press for our work as well as awards and recognition. Everything we did delighted everyone—except ourselves. We got to a point where we didn't want to just feed people. We wanted to end hunger. And that required an entirely different approach.

As we thought about how to pivot and achieve that goal we heeded the advice of social science writer Jonathan Kozol, who said that one should "pick battles big enough to matter, but small enough to win." The battle to end childhood hunger in the United States was just such a battle. Kids in America aren't hungry because there isn't enough food but because they lack access to public food and nutrition programs, and that's a solvable problem. But it meant that, like Brother Thomas, we had to break a lot of our own pots. We had to confront the notion that good was not good enough. It was not sufficient to please other nonprofits, reporters, politicians, or even funders. We had to achieve the best version of ourselves that we could be. We had to work differently.

After extensive research, we decided to hold ourselves accountable to the specific goal of ending childhood hunger by 2015. We invited many of our colleagues to join us. Not all of them were in favor of such a strategic change. At a meeting we convened of about fifty organizations, many of whom we worked closely with and funded, nearly all opposed the proposal at first. They raised questions about how we would measure progress, how we would fund such an effort, and what would happen if we failed.

Mostly, they were uncomfortable with being held accountable for specific, measurable, ambitious outcomes. Many had found satisfaction and rewards in doing good work and did not want to risk losing that. It took three years, but most eventually came around. Some of that was due to our persuasion, but much of it was because the times were changing. Other institutions were changing as well. President Obama adopted the goal of ending childhood hunger by 2015. Prominent governors asked Share Our Strength to bring our strategy to their states. Funders who had not previously supported Share Our Strength came to the table for the first time. We left some broken pottery on the ground, but that's something we all must challenge ourselves to do.

What we are attempting in our ongoing quest to end childhood hunger, and what Brother Thomas did with pottery, requires a shift in the way we think about what is possible. And it's the same shift that is occurring in the scientific labs of visionary malaria researchers. In this book, we will visit those labs and follow the work of one researcher, Stephen Hoffman of Sanaria labs, in particular. Hoffman and the vaccine he is relentlessly pursuing may be not only our best hope for eradicating malaria, but also our best modern example of how imagination, even in its most unreasonable forms—especially in its most unreasonable forms—can lead to breakthroughs.

This book is about more than these scientists—or even their work—because anyone who aspires to make a breakthrough and do what has never been done before can learn valuable lessons from their example.

MOST FAILURES ARE
FAILURES OF IMAGINATION

Whether in art, science, technology, or social activism, when there is failure, we often perceive and understand it as a failure of talent, strategy, planning, financial resources, or even execution. But those are not really the reasons most efforts fail. Most failures are failures of imagination. This is especially true for the seemingly intractable problems that have plagued us for decades, if not centuries. Albert Einstein said that "the specific problems we face cannot be solved using the same patterns of thought that were used to create them." Breaking out of those patterns demands a transformative, imaginative leap.

Examples of such triumphs of imagination are too few, but where they exist they are powerfully convincing.

The Institute for OneWorld Health in San Francisco is, more than anything else, a triumph of imagination by a former Food and Drug Administration official named Victoria Hale, who saw that a pharmaceutical firm could be structured as a nonprofit, released from the responsibility to maximize shareholder value, and made capable of accepting donated intellectual property from others. She essentially "took profit out of the equation" in developing and manufacturing medicines needed by the world's poor.[4] As a result, the institute, her brainchild, helps to actually create markets for drugs for neglected diseases. Established in 2000 as the first nonprofit pharmaceutical company in the United States, and now backed by more than $40 million from the Gates Foundation, it has created a new model for improving global health.

Teach for America, now the top employer of Ivy League graduates in the United States, was a triumph of imagination by a Princeton senior named Wendy Kopp in 1989. Kopp believed that the best students from the best universities would be willing to at least temporarily forgo careers in law and banking to teach in some of the most underserved schools in the country upon their graduation. There were countless obstacles to putting such a plan into action, ranging from the logistics of recruiting and training teachers to the resistance of teachers unions. But they were all surmounted by Kopp's imagination. Today, Teach for America has 7,300 current members teaching in thirty-five urban and rural areas. They impact 450,000 students annually, and nationwide there are more than 17,000 alums, including founders of charter schools, high-school principals, and school superintendents.[5]

The Harlem Children's Zone is a triumph of imagination by Geoffrey Canada, who conceived of the idea that some of the nation's poorest children should be surrounded, starting in utero, by a safety net woven so tightly that they would not be able to slip through it. Canada was president and CEO of a nonprofit called the Rheedlen Center, an organization that had been helping Harlem's children since 1970. But he was driven to do more, and in 1997 he launched a new initiative. By creating an interlocking network of services in a twenty-four-block area of Harlem, he wove that safety net, and the

Harlem Children's Zone was born. Children are testing at or above grade level on standardized tests and breaking the cycle of generational poverty as they graduate and enter the workforce. The effort has grown to encompass ninety-seven city blocks, and all the services are provided for free.[6]

Overcoming failure of imagination can be an enormous challenge. In some fields—including the nonprofit sector—the failure of imagination has become routine. In some ways, it is culturally ingrained thanks to severe and debilitating resource constraints. But imagination cannot be bought and installed like the latest software, or taught in an MBA program. Nor can it be inculcated into an organization by expensive consultants. There are no metrics by which it can be measured. That makes it easy to dismiss it as a "soft" resource, something that is "nice to have," rather than the "must have" hard currency that is needed to conquer seemingly intractable problems.

Though imagination cannot be purchased, there are ways to purposefully create a culture that acknowledges the primacy of imagination in reaching breakthrough solutions. This can be done by constantly challenging the conventional wisdom and even the most longstanding assumptions. It can be done by asking hard questions about what is possible, even if such questions seem naïve, and by rewarding risk and not penalizing dreamers.

Imagination can be nurtured and elevated by properly funding R&D—which is often considered a luxury—as if it were a necessity, because it is. And it can be stimulated by forcing those in an organization, from the senior leadership on down, to get out from behind their desks and venture into places where their imaginations will be stimulated by bearing witness to people and places very different from themselves.

Sun Tzu, the ancient Chinese military strategist who wrote *The Art of War*, said that every battle is won or lost before it is fought.[7] Similarly, every effort to change the world journeys toward its destiny on a path determined by what can be imagined. Not believing that we could end childhood hunger was a failure of imagination, and it distorted and undermined the way in which the anti-hunger community went about its work for generations. Not believing that malaria could actually be eradicated was a failure of imagination that distorted and undermined the way the malaria community went about its work, until someone leaped unreasonably over the hurdle.

IRRATIONAL CONFIDENCE: THE VISIONARY'S DILEMMA

These two strands of belief—that good is not good enough, and that most failures are failures of imagination—when woven together, are held fast by the glue of unshakable belief in oneself. It is old-fashioned advice, this notion of believing in oneself, the stuff of commencement speeches and testimonial dinners. But in certain circumstances, continuing to believe in yourself and your calling, even against all odds, can be determinative. The bigger one's dreams, the more tangible and important such belief becomes.

The visionary's dilemma is that the bigger the goal or aspiration, the bolder and more audacious the plan for attaining it, and the more skeptics and cynics there will be. The dilemma is particularly pernicious because it persists and compounds. The more the visionary pushes and pursues, the more the establishment interprets this as a sign of fundamental instability, conveniently justifying its initial opposition. Concerns about the idea are compounded by concerns about the idea's propagator. Establishments are threatened by visionaries, especially when, as often happens, a visionary's approach suggests that the solution has been hiding in plain sight all along, notwithstanding the phalanxes of bright people who have dedicated their entire careers to more conventional approaches.

The status quo yields not an inch of ground without a fight. The establishment always has the advantage of money, credibility, respect, prestige, familiarity, and political support. Just as a daring quarterback's consistent effectiveness all but invites the defensive line to blitz, the visionary has to expect the pass rush and hold his or her ground.

So a visionary's best defense to the dilemma is not only having a thick skin, but having reservoirs of self-confidence as well. Because when those invested in the status quo feel threatened, they chip away at not only the upstart's ideas, but also his or her motives and character. Just as big trucks require big wheels, and tall buildings require deep foundations, people with big dreams need a large reservoir of self-confidence to maintain their balance and go forward. It helps if friends and family can be depended upon to help fill it.

I don't think it's an accident that many of the people profiled in the story of the pursuit of a malaria vaccine are directly supported by family members such as spouses and sons and daughters. The original discovery of a potential malaria vaccine in 1968 was by the husband-and-wife team of Victor and Ruth Nussenzweig, who even today, in their eighties, share a lab at New York University. Their discovery was the genesis of future work pursued by several other husband-and-wife teams, including Steve Hoffman and Kim Lee Sim in the United States, as well as Pedro Alonso and his wife Clara Menendez in Spain and Africa. Inherent in such familial couplings is a support system, a kind of anchor that helps such people weather the inevitable storms. Standing alone against the multitudes requires a degree of belief in oneself that simply surpasses the rational.

THE IMAGINATIONS OF UNREASONABLE MEN

The three philosophical underpinnings of breakthrough thinking described above—(1) that good is not good enough, (2) that most failures are failures of imagination, and (3) that irrational self-confidence is essential—are not by themselves a solution for our toughest problems. They are not even a shortcut to such solutions. But they are the necessary architecture for solving them, the underpinnings without which most efforts will falter.

Just look around. We are surrounded by the monuments of men and women who failed to recognize the stop signs along their journey to solving a problem or creating something new.

As Dan Pallotta, founder of the ambitious and wildly successful AIDSRides, bicycle rides to raise funds for AIDS service organizations, once said to me: "Don't you suppose someone must have argued to Henry Ford: 'But that's crazy—you'd have to build these gas station places all over the country and pave these incredibly long roads.'" Great imaginations are almost always unreasonable, but they almost always triumph in the end.

Most of us won't cure malaria or invent the next automobile. So why are these elements of breakthrough thinking important in our own lives? Can they apply to each of us? They do if we believe that the organizations, communities, and world of which we are a part can do better. They are important if we're frustrated with the slow and incremental pace of social change, or if we wish to play some small role in lightening the suffering and struggles of those less fortunate with whom we share this planet. They are the qualities that allow some people, gifted with great vision, to insist that, rather than taking the reasonable approach of adapting to the world, the world, in George Bernard Shaw's words, must adapt itself to the unreasonable man.

CHAPTER 2

WHATEVER IT TAKES

Researchers at Edinburgh University's Institute of Cell and Molecular Biology have isolated part of a protein which allows [malaria] to become resistant to new treatments. . . .

Professor Malcolm Walkinshaw, of Edinburgh University, said: "We can now use this protein structure to design a new generation of drugs which makes it possible to overcome resistant strains of malaria.

"People have studied this protein for a long time, but until now, no one has been able to determine its detailed structure. This is a real breakthrough."

—BBC News, "Malaria Treatment Breakthrough,"
April 22, 2003

STEVE HOFFMAN IS A DOCTOR who wants to develop a vaccine to prevent malaria. If it works it will save the lives of millions of children. If it doesn't, he will find company in the ranks of countless others who have gone before him, tried, and failed. And millions will continue to perish in an agonizing death.

While there are vaccines for bacteria and viruses, there has never been a vaccine for malaria, or for any parasitic disease.

The reasons are both scientific and political. The parasite is complex, elusive, and even brilliant, in an evolutionary sense. And most of those whom it infects are so voiceless, vulnerable, and marginalized that there are no markets—economic or political—for serving them or solving the problems they face. They are victims not only of malaria but also of chronic political laryngitis. And their condition persists not because of the paucity of solutions, but because they have no political voice. Society has not been fully persuaded to pay for solutions that already exist. Nor are there many who are willing to share in the sacrifice of time and money that would be required to sustain those solutions and take them to scale.

Instead their fate depends not on traditional approaches but on an emerging new cocktail of entrepreneurship, philanthropy, and science—stirred by imagination—a cocktail being developed by Steve Hoffman and a handful of colleagues and competitors around the globe.

AN UNSTOPPABLE FORCE

Steve Hoffman has practiced medicine for more than twenty-five years, but his ambition has required him to wear many hats: that of naval officer, business entrepreneur, research scientist, author, evangelist, employer, fundraiser, humanitarian, and biotech engineer. He has the necessary personality traits to go with them: brilliant, innovative, confident, arrogant, impatient, stubborn, charming, abrasive, driven, determined, and competitive.

One particular quality dominates all of the others, giving Hoffman's quest its unstoppable force: He is prepared to do whatever it takes to save millions of children's lives.

When, in 1980, it meant leaving a comfortable medical practice in San Diego and joining the U.S. Navy in exchange for access to the best research tools, labs, and clinics, a young Dr. Hoffman cut his hair, shaved his beard, and enlisted, shipping out to Indonesia, which at the time hosted our government's most advanced infectious disease research lab in the tropics.

When, in 1987, it meant testing the safety and effectiveness of a potential vaccine he'd helped develop, Hoffman rolled up his sleeves—or, literally, a sleeve—and let a swarm of mosquitoes drink blood from his arm until they transmitted in their saliva enough deadly parasites to make their way to his liver. That vaccine failed and he proceeded to get violently ill with the chills, shakes, and fever of malaria.

When it meant starting a private biotech company to pursue a vaccine-development approach that every other expert dismissed as impossible, he retired as a navy captain and, in 2003, started the business with his wife and son at their kitchen table, raising the capital, hiring the technicians, and quietly constructing one of the most unusual laboratories in the world to extract microscopic parasites from the dissected salivary glands of live malaria-infected mosquitoes.

When it meant establishing his authority as an expert with whom others would collaborate and invest, he experimented, researched, and wrote more than 350 papers over the years, becoming the most cited author in the world for scientific articles on malaria.

And when it meant convincing a skeptical scientific community that there was a viable alternative to the better-known vaccine being tested by GlaxoSmithKline, the fourth-largest pharmaceutical company in the world, he packed up his PowerPoint slides and flew to international conferences in Dakar, London, Nairobi, Amsterdam, and countless others cities to stand alone and make the case for his unusual approach.[1]

Whatever it takes.

But whenever one mentioned Steve Hoffman's name to his colleagues, those who were accepted as members of the small, dedicated fraternity of tropical disease specialists who had also dedicated their lives and careers to battling malaria, you'd get the kind of silence, stare, or stutter that prompted you to change topics. Eyebrows arched. Heads shook. The skepticism was palpable. Nevertheless, it was leavened with respect and sometimes envy. When someone is willing to do whatever it takes, you never quite discount him. I, for one, was incredibly intrigued.

I first heard Steve Hoffman's name in 2005 when it was mentioned to me by Dr. Diane Griffin, who chaired the Department of Molecular Microbiology and Immunology at Johns Hopkins University's Bloomberg School of Public Health, the largest school of public health in the world, named for its largest donor, New York businessman and mayor Michael Bloomberg. The Oklahoma-born Griffin had been lured there from Stanford University, along with her husband, in 1970. She had become an assistant professor in 1973, an associate professor in 1986, and, eight years later, chair of the department. When she spoke of Hoffman, she almost giggled. That was when I knew I had to find out what he was up to.

"AND THEN THERE'S THIS
WILD THING STEVE HOFFMAN IS DOING"

Griffin's own specialty has been the measles virus. In May 2001, though, the Bloomberg School received a $100 million gift to be used in the battle against malaria. Dr. Griffin, a well-respected veteran researcher on how viruses create infectious disease, with 275 published papers of her own, was tapped to establish a Malaria Research Institute and to map a strategy and attract the talent that would make it distinctive.

Her work on measles has been impressive. In America, vaccinations keep us safe from measles, but the disease still takes the lives of thousands of children in developing countries. It was Griffin who discovered that the measles virus could suppress the release of the protein Interleuken 12, weakening the body's natural immunities. This effect makes measles especially dangerous in developing countries because it leaves children vulnerable to other infectious agents, such as pneumonia and malaria. Her research has led to work on several promising vaccines.

"I've always been interested in problems that affect those without a voice," she explained. "Malaria is very different from AIDS, which affects adults who can organize themselves and lobby." She is particularly focused on how the fight against malaria has been fought to a stalemate:

> The vaccine efforts around malaria haven't changed for thirty years. They need new input. Even the diagnostic blood smear is the same. And more children are dying from malaria now than died ten years ago. . . .
>
> We believe that there is no magic bullet, and therefore when we got the grant we decided to recruit broadly and tackle the issue at all points. We've developed a somewhat unique emphasis on the mosquito, rather than the parasite. Historically, success has always been a result of controlling the means of transmission, which we call the vector. At our first international conference on malaria, which was dominated by vaccine people, many had been so focused on the parasite that they had not heard talk of mosquitoes before.[2]

Dr. Griffin generously offered to set up a meeting for me with "the mosquito people," a team that was working on a vaccine to prevent transmission of malaria from mosquito to mosquito. "The only progress that's ever been made with malaria has been in controlling its vector," she said. If mosquitoes can't transmit malaria to each other, then transmission to humans begins to decline also.

Finally, almost in passing, she mused, "And then there's this wild thing Steve Hoffman is doing. He used to run the navy's program and apparently has all of these people out in a lab somewhere, all bent over their desks dissecting live mosquitoes." I could see from the girlish grin she flashed and the way her eyes lit up that she found an element of eccentricity in what Hoffman was doing. And while she said no more, I also sensed she wouldn't have mentioned it if she didn't know and respect Hoffman.

In a field that has suffered a stalemate for many years, eccentricity might be just what is needed. Hoffman sounded like renegade of sorts. I guessed that he'd been driven by urgency and frustration to journey from the establishment figure he once was—with its uniforms, salutes, and rulebooks—to counterculture rebel.

I made a mental note of Hoffman's name. When I got home I discovered that his medical and scientific credentials were impeccable. Educated at the University of Pennsylvania and Cornell and with a diploma from the prestigious London School of Hygiene and Tropical Medicine, Hoffman became a captain in the navy, rising to direct the malaria program at the Naval Medical Research Center, and coordinated the Department of Defense's malaria vaccine development efforts. He had 340 scientific publications to his credit, had received the navy's most prestigious award for scientific achievement, and had served as president of the American Society of Tropical Medicine and Hygiene. He'd been recruited by Craig Venter, the maverick biologist who raced the U.S. government to sequence the human genome, to become a senior vice president at Celera Genomics.

Having risen to the very top of his field, he then walked away from it all—from the opportunities for financial gain at an industry leader like Celera, and from the resources and security of a large and powerful institution like the U.S. Navy. Instead he was now laboring away in near obscurity— and as the occasional target of ridicule and scorn—in a small lab he had cobbled together from scratch. I couldn't think of anyone I'd rather meet.

CHAPTER 3

STRANGE BEDFELLOWS

A team of Monash University researchers . . . has made a major break-through in the international fight against malaria. . . .

The team . . . has been able to deactivate the final stage of the malaria parasite's digestive machinery, effectively starving the parasite of nutrients and disabling its survival mechanism.

—*Science Daily*, "Breakthrough to Treat Malaria,"
February 8, 2009

IN 1943, GENERAL DOUGLAS MACARTHUR, the U.S. commander in the Pacific, said: "This will be a long war if for every division I have facing the enemy I must count on a second division in hospital with malaria and a third division convalescing from this debilitating disease."[1] During the Solomon Islands campaign, malaria caused more casualties than Japanese bullets, and after the initial landings on Guadalcanal, the number of patients hospitalized with malaria exceeded all other diseases. Some units suffered 100 percent infection rates, with personnel sometimes being hospitalized more than once. Though we were ultimately victorious over the relentless Japanese military, another even more daunting foe—the malaria parasite—remained undefeated.

In fact, malaria was a prime concern of military commanders the world over long before World War II. Alexander the Great is believed to have been killed by malaria at the height of his powers. The disease wiped out the army of France's Henry II in the 1500s. One of the first expenditures of the Continental Congress during the American Revolutionary War was $300 to buy quinine to protect General Washington's troops. Ten thousand soldiers died from the disease during the U.S. Civil War, and there were at least 600,000 cases, primarily in the South Pacific, during World War II. In some areas of the South Pacific, malaria rates were four cases per person per year. The end of our campaign in the Pacific marked the beginning of escalating combat against this dangerous killer.[2]

Malaria causes flu-like symptoms: chills, headaches, nausea, and vomiting. It can escalate, at any age, into anemia and punishing fevers, convulsions, and coma, but is fatal mostly to children under five. Kidney failure, a ruptured spleen, pulmonary edema (fluid buildup in the lungs), cardiovascular shock, and collapse are all potential outcomes.

Its most dangerous form is cerebral malaria, in which the blood vessels carrying blood to the brain become clogged. Red blood cells that have been invaded by parasites develop knobs on their surfaces, which enables them to cling and stick to the capillaries and small blood vessels. The parasite means no harm—it's merely trying to avoid being sucked into the filter of the spleen, whose job it is to weed out and destroy damaged red blood cells. Left untreated, cerebral malaria destroys blood vessels in the brain and is fatal in twenty-four to seventy-two hours.

The surest way to know whether you have malaria is to have a diagnostic test where a drop of your blood is examined under a microscope for the presence of malaria parasites. Even this simple and reliable procedure is problematic in Africa, where cultural issues create a reluctance to give blood. (Scientists at the Johns Hopkins Malaria Research Institute are experimenting with a urine dipstick test that they hope will have greater utility.)

For hundreds of years, quinine was used to treat the effects of malaria. Quinine allowed the body to build up a substance called *heme* that was toxic to the malaria parasite and could clear it from one's system. It was not used to eradicate the disease but to mitigate its effects in anyone who contracted it. Quinine was derived from the bark of the cinchona tree, originally confined to the Andes Mountains of South America, but smuggled to other countries because of its curative powers. The tree is now found in Bolivia, Java, and India, among other places. Like most natural remedies, the supply is constrained, which increases the price. There are forty species of cinchona, which was named for a Peruvian countess. One accesses its curative powers by beating the tree trunks and then removing the peeling bark. One or two grams of ground or chopped bark boiled in water yield the quinine, which is an alkaloid chemical. Its active ingredients were farmed and manufactured synthetically.

When the supply from South Pacific countries was cut off by a Japanese military blockade in 1941, the U.S. military began to focus on drug research and development. In 1934, research by German scientists to discover a substitute for quinine led to the synthesis of sontochin. During World War II, French soldiers happened upon a stash of German-manufactured sontochin in Tunis and handed it over to the Americans. American researchers made slight adjustments to the captured drug to enhance its efficacy. The new formulation was called chloroquine. Because it could be synthesized in the lab and therefore cheaply mass-produced, chloroquine revolutionized treatment of malaria, pushing quinine to the sidelines.

By the 1960s, resistance to chloroquine had developed and spread. Mefloquine was then created. Mefloquine kills parasites once they enter red blood cells and stops them from multiplying further. But no one is exactly sure how it does so. The prevailing theory is that, like chloroquine, it works by blocking the action of a chemical that the parasite produces to protect itself. The parasite digests hemoglobin and keeps the globin, but heme is toxic to it, so it produces a chemical that converts the heme into a harmless compound. It's like eating blue crabs and keeping the meat but throwing away the awful-tasting gills.

Lariam—the brand name under which mefloquine is sold—is strong stuff. Its influence is felt beyond the parasite it is designed to kill. Its side effects range from severe depression and paranoia to vivid dreams. So, in addition to its expense, its toxicity makes it impractical as a long-term solution. It is also a drug that works only to mitigate the disease. It is only a temporary substitute for a vaccine.

In 1941 our armed services began collaborating with universities and pharmaceutical companies and produced a new arsenal of effective synthetic alternatives. But during the Vietnam War, resistance to the drugs emerged. In 1963 the U.S. Army began a new program that produced two dozen more antimalarial drugs within eleven years. Still, in Vietnam the disease reduced some combat units by half, and it became ever more apparent that vaccine development was a worthwhile investment.

MOSQUITOES MORE DANGEROUS
THAN MORTAR ROUNDS

Malaria creates the strangest bedfellows. The two most likely victims occupy opposite extremes on humanity's spectrum: the poorest, weakest, least educated, and most vulnerable children on the planet, on the one hand, and, on the other, the strongest, best-supplied, most magnificently trained, healthiest soldiers in the world, who make up America's military forces and are often deployed to regions where malaria is endemic. These two groups are different from each other in almost every imaginable way, but they're both vulnerable to a parasite that doesn't discriminate.

Between 300 million and 500 million people are infected with malaria each year. Adults usually survive and, though sick and weakened, develop immunity from its fatal form. While childhood mortality from all causes is decreasing in Africa, malaria mortality is actually increasing. The economic toll is believed to be billions of dollars a year in Africa. Malaria exists in eighty-one countries. Health organizations estimate that up to 5 million people have died of AIDS over the past fifteen years. During the same time period, nearly 50 million have died of malaria.[3]

When I traveled to Ethiopia in 2002 on the trip during which I met Alima, and the time before that as well, there was no question about whether I would be protected from malaria, not to mention a variety of other tropical diseases. Though I would be jet-lagged and exposed to the elements, staying in places without insecticide-treated bed nets, and lacking knowledge of the terrain and the local ecosystem, I had access to something far more valuable: a Travelers Clinic just a few blocks from my office in downtown Washington, D.C. And I had the $350—more than twice the average annual income of an Ethiopian—to purchase prophylactic medicines.

There is nothing out of the ordinary about this if you are from a modern country in the developed world. In fact, what would be extraordinary, even newsworthy, would be if I had returned with malaria. This happens, but it is extremely rare. So why are some of us at minimal risk and others constantly flirting with death from a disease we know how to cure?

Access to drugs and vaccines is only part of the answer. The rest has to do with access to even the most basic medical care. The United States has approximately one physician for every 500 people. Ethiopia has one for every 36,000. The population of Ethiopia is 70 million. There are more Ethiopian doctors in Chicago than there are in Ethiopia. Because many doctors are based where the population is concentrated, there are many rural areas where the ratio would be much worse than one in 36,000.[4]

The first time I visited a field hospital in one of the famine-struck regions of the country, it was an hour before I even realized I was in the "hospital." It was an open-air tent with about sixty ragged straw mats on the ground, a baby laying listlessly on each one, and their mothers beside them, each mom either trying to get some food, water, or medicine into her child or taking a rare opportunity to sleep herself. The mothers covered most of the duties that we expect of nurses. There was a doctor, but virtually no sign of medical equipment, monitors, or any of the paraphernalia one normally associates with a hospital.

The doctor's ability to diagnose was seriously compromised; his ability to treat an illness, once diagnosed, even more so. Comfort was not even a consideration. The likelihood of keeping such a facility supplied with the right drugs—the likelihood of drugs reaching such a remote spot on a regular basis—seemed low. If there was anywhere that the moral, scientific, and economic arguments for a vaccine converged, I thought, it was under such a tent.

Dr. Denise Doolan, who was scientific director of the malaria program at the U.S. Naval Medical Research Center, has said, "In all conflicts in the past century in malaria endemic areas, malaria has been the leading cause of casualties, exceeding enemy-inflicted casualties in its impact on person days lost from duty." According to a doctor at the U.S. Army research lab, the disease is considered "the primary factor degrading combat efficiency" in those regions.[5]

In a 2005 article published in the *Naval War College Review* entitled "The Mosquito Can Be More Dangerous Than the Mortar Round," authors Craig Smith and Arthur Hooper documented why disease and illness will likely generate more casualties than combat during military operations in either Africa or Asia. Data from Vietnam alone showed that only a third of hospital admissions were from combat wounds. Two-thirds were from disease and non-battle injuries.[6]

Using the 26th Marine Expeditionary Unit's insertion into Liberia in 2003 as a case study, the authors reported that 80 of 230 troops experienced symptoms of malaria and that "the outbreak was a blow to combat effectiveness." Several victims developed cerebral malaria, in which blood vessels carrying blood to the brain became clogged.[7]

Military history is rife with such examples. In 1942, the First Marine Division was pulled from combat and sent to

Melbourne to recuperate because 10,000 of its 17,000 men were incapacitated by malaria. The disease hit 85 percent among the men holding onto Bataan. As the commander of British forces in Burma during World War II, Field Marshall Archibald Wavell, wrote: "We must be prepared to meet malaria by training as strict and earnest as that against enemy troops. We must be as practiced in our weapons against it as we are with a rifle."[8]

Smith and Hooper concluded that "these realities could easily render a U.S. military force ineffective without a combat engagement ever taking place."[9]

FROM IMMUNE
RESPONSE TO VACCINE

Steve Hoffman and Rip Ballou started out in the mid-1980s as friends, colleagues, and allies committed to developing a malaria vaccine that would protect soldiers and children alike. From very different backgrounds, they had both found their way to military commands in Washington, D.C., at about the same time and the same age.

Ballou, a fourth-generation army officer born at Fort Campbell, Kentucky, had dropped out of West Point after one year but ended up making a four-year commitment to the army as a way of financing his education at Georgia Tech. He then got his medical degree at Emory. In the early 1980s he began work in tropical diseases at the Walter Reed Army Institute of Research.

Hoffman had shifted from political science to pre-med during an Ivy League education at the University of Pennsylvania. After getting a medical degree from Cornell he joined the navy to practice tropical medicine and spent four and a half years in Jakarta before returning to the United States in 1984. While there, he worked on severe typhoid fever, and *The New England Journal of Medicine* article he authored that year set the standard for treating the disease. He considers it his most important contribution to date. For the next sixteen years he would lead the malaria program at the Naval Medical Research Center.

Before long, Hoffman and Ballou had teamed up to try to build on the most significant breakthrough in malaria-vaccine development of the time, which had come from New York University researchers Ruth Nussenzweig and Jerome Vandenberg.

In 1967, Nussenzweig, building upon earlier work by Vandenberg, had shown that it was possible to prevent malaria infection in mice by immunizing them with irradiated parasites. Unlike quinine or chloroquine used to treat malaria, irradiated parasites could actually alert and trigger one's immune system in advance, as vaccines are designed to do, to prevent infection in the first place. But Nussenzweig acknowledged that proving the value of irradiated parasites as an immunogen was quite different from proving that they could serve as an effective vaccine. After all, such parasites could be harvested from only one place—the salivary glands of infected mosquitoes. How could that ever be a reliable supply for the massive quantities needed? And they had been introduced as an immunogen in only one way—through the bites of infected mosquitoes. No one had ever even experimented with introducing such an immunogen into human bodies. Of the many possibilities, at least one could probably be counted out: soldiers, business travelers, tourists, and impoverished children rolling up their sleeves and allowing a swarm of hungry mosquitoes trapped in a canister to feed on their blood.

Ruth Nussenzweig still recalls that when she first arrived at NYU, "scientists thought a malaria vaccine was impossible" because the many stages of the parasite's life cycle made it difficult to target one part for use in a vaccine. The first trials of their vaccines were with volunteer prisoners at a maximum-security facility in Jessup, Maryland. According to the *Baltimore Sun*, a Baltimore longshoreman who killed a man in a bar fight might have been the first human ever immunized against malaria. But letting malaria-infected mosquitoes that had been exposed to X-rays feed on inmate volunteers was still a long way from a practical vaccine.[10]

In 1980, Ruth Nussenzweig and her husband, Victor, became the first to identify and isolate a protein that coats the malaria parasite. It was called the *circumsporozoite protein*, or CS protein, and they were also the first to show that it was possible to generate antibodies for an immune response against it.

Hoffman and Ballou set out together to turn that immune response into a vaccine, and their effort became the basis for a vaccine candidate. The army and navy worked on it together, and it was ultimately adopted by Glaxo-SmithKline, funded by the Gates Foundation, and put to one of the largest clinical trials ever conducted in the developing world.

Today that vaccine is known as RTS,S, and for two decades it was on a rollercoaster ride of small victories followed by dashed expectations.

FROM COLLABORATORS
TO COMPETITORS

In 1987, when Hoffman was at the Naval Medical Research Institute and Ballou at Walter Reed Army Institute of Research, they were so confident in the vaccine they'd developed together that they vaccinated themselves and then challenged a small group of four other colleagues, who agreed to volunteer. Their confidence was misplaced. Within a few weeks, five of the six became violently ill with malarial fevers, Steve after flying cross-country to give a presentation in San Diego, Rip after running six miles and drinking a beer. Ballou said the ensuing headache felt like "a 9-inch spike through my head."[11]

The vaccine was then refined by combining a protein from a parasite with another from the hepatitis-B virus. It worked in two of eight volunteers exposed to malaria. Then an adjuvant, or chemical additive that boosts the body's immune response, was included, and the RTS,S vaccine was given the commercial name Mosquirix. A 1996 test at Walter Reed showed its promise. The shot reduced new malaria infections in the first two to three weeks by 86 percent, but after that short period it only reduced infection by 30 to 40 percent. Hoffman and others judged it a good breakthrough—but not good enough.

Hoffman and Ballou began to diverge in their approaches, and neither had an easy time transforming his vision into reality. Ballou later told *Scientific American* that "it turns out that mice are very easy to protect, and humans are not." Hoffman, describing his radical effort to extract the weakened parasites that would be used as the vaccine from the salivary glands of irradiated mosquitoes, told *Business 2.0*: "It's not a hard process unless you try to do it."[12]

Back when they'd joined the military, it was the only game in town for working on tropical disease and vaccine development. No one else had the money, the scientific facilities, or the pressing need—represented by half a century of disease-related military casualties from Guadalcanal to Liberia—that would justify such investment. But by 2000, revolutions in molecular biology, biotech, and genomics, along with new sources of philanthropic funding, had changed all that. There were private-sector options for both Hoffman and Ballou that simply hadn't existed before. They seized them.

Within two years of each other Hoffman and Ballou left the military and went into the private sector to pursue their efforts. Hoffman joined Craig Venter at Celera in another type of race—to map the human genome and find cures for cancer. Hoffman described Celera as "a company with vision, courage, and a billion dollars in the bank." While there, he talked Venter into mapping the mosquito genome. When Venter and his board clashed over the direction of the company, Venter left. Hoffman soon left, too.

Ballou retired from the army in 1999 and spent the next eight years in the vaccine industry, including five years at GlaxoSmithKline Biologicals in Belgium, where he was responsible for the company's clinical development programs for malaria vaccines. In April 2008, he left GSK to join the Bill and Melinda Gates Foundation as the deputy director for vaccines, infectious diseases development, in the Global Health Division. Ironically, Ballou would be in a position at Gates to influence decisions about funding for Hoffman's work. In 2009 he returned to GSK.

Hoffman's path could not have been more different from Ballou's. He didn't join a company but started one, creating Sanaria at his kitchen table. While Ballou was ensconced on GlaxoSmithKline's vast Belgian campus, tapping into the expertise of thousands of scientists, Hoffman was cobbling together a start-up with his wife Kim Lee and his three sons. Dissatisfied with the efficacy of RTS,S—"That's not a vaccine that could ever be considered for use in the developed world"—Hoffman returned to the Nussenzweig discovery regarding the potential of the weakened parasite to serve as a vaccine, obsessed with the fact that it actually worked, even if no one could figure out how it could be made.

At that kitchen table Hoffman set out on a path considered so impractical, so unreasonable, that even the Nussenzweigs held no hope that his goals could be achieved. Build a lab that could breed mosquitoes, and dissect their salivary glands in search of sufficient quantities of parasites to inoculate hundreds of millions of children on the other side of the planet? The question was not what could go wrong but what wouldn't. Growing sufficient parasites? Harvesting them sterile and pure from the salivary gland, of all places? Keeping this "vaccine" stable and effective as it was transported thousands of miles to Africa or Asia? And delivering it how—by shot? By pill? By the bite of mosquitoes?

But Hoffman didn't have a boss or board to stop him, or a bureaucracy to slow him down. For the first time in his career he would report to no one but himself, be accountable to no vision or ambition other than his own.

Unlike the time he spent working with Venter at Celera, though, this time there was no billion-dollar bankroll. The venture required classic bootstrapping, one small step at a time, each glimmer of progress parlayed into a hopeful headline that would yield more funding and more glimmers.

Though in the beginning Hoffman and Ballou were more collaborators than competitors, as the stakes grew higher their rivalry grew sharper and ultimately fierce. Tens of millions of dollars had already been invested in and by each scientist. Hundreds of millions of dollars more were on the table. Neither was growing younger. Each had something to prove. Though they were no longer in the military, it was soon hand-to-hand combat all over again.

Each sought to turn his adversary's strengths against him and to turn his own weaknesses into assets. In contrast to GlaxoSmithKline's enormous bureaucracy, Hoffman could be agile, take risks, and work outside of the glare of the spotlight. GSK, on the other hand, could use its size, money, and media machine to maintain momentum even in the face of less-than-compelling results.

For all of GSK's enormous advantages in resources, Hoffman maintained one leading edge. Large corporate institutions like GSK are not usually the best places to foster imagination or creativity. As William Deresiewicz, a former Yale professor and a widely published literary critic, cautioned the incoming plebes in a lecture at West Point in October 2009:

> We have a crisis of leadership in this country, in every institution . . .
>
> . . . And for too long we have been training leaders who only know how to keep the routine going. Who can answer questions, but don't know how to ask them. Who can fulfill goals, but don't know how to set them. Who think about *how* to get things done, but not whether they're worth doing in the first place. What we *don't* have are leaders.
>
> What we don't have, in other words, are *thinkers*. People who can think for themselves. People who can formulate a new direction: for the country, for a corporation or a col-

lege, for the Army—a new way of doing things, a new way of looking at things. People, in other words, with *vision*.[13]

The implication is more than a little unfair to Rip Ballou, a good and dedicated man working with thousands of talented and committed people, and actually saving lives. But though unfair, it is not invalid. Hoffman has a way of looking at the same things everyone else is looking at and seeing something different. I sometimes have wondered how he suffered the hierarchy of the military for so many years, or how they suffered him. But it's no accident that a man of such imagination, a man for whom good is not good enough, is achieving his greatest success while not part of some bureaucracy, whether military, corporate, or governmental.

CHAPTER 4

THE BIO-HAZARD
LEVEL 3 STRIP MALL

Thanks to . . . U.S. and Danish researchers the world might be free of malaria soon. Finding a solution to nip the problem in the bud, the researchers have developed a way to attack the gene . . . [that] helps the malarial parasite to reproduce inside the mosquitoes. . . .

The scientists believe it to be a major breakthrough, because, if the reproduction of parasites is stopped[,] the spread of malaria via mosquito bite can be checked.

—Themedguru.com, "Attacking Malaria by Nipping
It in the Bud—Study," June 4, 2008

IN FEBRUARY 2006, WHEN I first made plans to visit Steve Hoffman at Sanaria, I expected to find him in a shiny high-tech complex or an industrial park, perhaps on a sprawling campus. Instead, Mapquest turned up a small strip mall in Rockville, just a few minutes from where I used to live in Silver Spring, Maryland. It was in a neighborhood of nondescript self-storage centers, U-Haul lots, and home-furnishing suppliers.

Sanaria was the company he'd formed in 2003, the realization of his dreams for malaria research that he'd first shared with his family at his kitchen table. In August 2007, he moved its headquarters to a much more advanced facility, also in Rockville, where it is part of Maryland's Biotechnology Corridor.[1]

Sanaria has come a long way. When I visited the original site, Suite L on the second floor of 12115 Park Lawn Drive in Keats Plaza, the company's closest neighbors included a Floor Covering Center and Tri-Graphics Picture Framing. A glass door with small white stick-on letters spelling the company name confirmed that I was in the right place, but from the outside it looked like the kind of space one would rent for a temporary project, maybe a campaign headquarters for county commissioner or an interim sales office. If this was the impression the neighbors took away, I thought, that was just fine. If they'd had the slightest inkling of what they were driving by each day—of what a mild breeze might send toward their doorstep—Steve Hoffman might have had his hands full with a lot more than science and biotechnology.

Having anticipated test tubes, white lab coats, and state-of-the-art equipment, I felt disappointed at the idea of talking to someone at a desk in this shabby building. The glass door was locked, and I leaned my nose against it to peer in, wondering if anyone was even there. A woman soon appeared, opened the door, and said Dr. Hoffman would be with me shortly.

She did not have to usher me through a maze of corridors into some inner sanctum to see her boss. Although Hoffman was president of the company, his office was right by the front door (as I found out later, the inner sanctum was reserved for the mosquitoes). There was no place for me to sit, so I just stood there waiting, next to a strange neon blue light that hung down in a wire cage near the corner of the ceiling.

"Three hundred and nine thousand!" a voice shouted from the other side of a thin plywood wall. "That's right, 309,000, can you believe it? And that's just one. Altogether we did 109 million just today." I could tell that Hoffman was on the phone. There was exuberance in his voice, and confidence, even a trace of mischief. Then I heard other voices and realized he had colleagues in his office. They were sharing in a moment of triumph, celebrating it. But it was his voice that was dominant.

If we'd been in New York City, I'd have thought he'd made a killing in the stock market that very afternoon. "A hundred and nine million!" he shouted again, even louder, as if it were shares or dollars.

"You came on a banner day, probably the best day we've ever had," he announced a few moments later. "Right in there," he said, pointing back to a locked door with universally recognized bio-hazard warning symbols on it, "right there, this afternoon, we dissected the salivary glands of enough mosquitoes to extract 109 million parasites."

He wasn't speaking of just any parasite, but of *Plasmodium falciparum*, the parasite that causes the deadliest form of malaria. It kills more children—1 million every year—than any other single infectious agent on the planet and is responsible for untold suffering in dozens of countries. It is the U.S. Defense Department's number one science priority after terrorism, though hundreds of billions of dollars are spent on the latter and only tens of millions on the former. It has been attacked with every strategy known to man, and it has never been beaten.

With evolution and natural selection on its side, *P. falciparum* has prevailed over every effort to destroy it. Pitted against the greatest minds, massive resources, and the most evolved technology, beginning with the microscope and through to computer mapping of the human genome, the single-celled parasite has consistently outwitted and outmaneuvered its foes at every turn, somehow managing to keep concealed nature's deepest and most mysterious secrets. It is a battle that has been truly epic in proportion.

Listening to Hoffman, I was momentarily distracted by trying to call up a mental image of what it takes to dissect a mosquito's salivary gland. I pictured small technicians in white coats hunched over large microscopes at flat white tables. My vision was not far off, but it turns out that dissection is the easy part. Breeding the mosquitoes in an environment sterile enough for approval by the Food and Drug Administration, ensuring the vaccine could be stored safely and would remain chemically stable, testing to rule out all possible side effects—these were only a few of the formidable challenges Hoffman faced beyond extracting sufficient quantities of parasite. The process was memorably described by Jason Fagone writing in *Esquire*: "Hoffman's vaccine would have to be made inside mosquitoes. It would be like baking a pie in a cow."[2]

My science education was just beginning; I still had a lot to learn. But I knew enough about Hoffman's vaccine theory to know that this milestone of harvesting 109 million parasites in one day was a critical development. It affirmed that he'd eventually have enough raw material to produce the vaccine at the scale required.

Although a vaccine would be the ideal way stop the *P. falciparum* parasite and to prevent malaria, there is not now, and never has been, a licensed malaria vaccine. In fact, there has never been a vaccine for any parasitic disease. The vaccines we are familiar with, for polio, measles, and smallpox, to name a few, are for viruses.

Sanaria is the only company in the world dedicated entirely to malaria vaccine development. Steve Hoffman is quick to point out its promise. The immunogen—that is, the specific set of proteins that prompts the body to fight back against foreign invaders—that Sanaria was using for its vaccine, he explained,

has been shown to protect 13 of 14 (93 percent) of human volunteers against 33 of 35 (94 percent) experimental infections for at least 10 months. No other experimental

malaria vaccine has ever been shown to consistently protect more than 40 percent of experimentally infected volunteers for more than three weeks, and no competitor is
working on an experimental malaria vaccine with the potential to protect even 50 percent of volunteers against infection for more than a few weeks.

The ever-resilient, mutating parasite has prevailed over
every effort to stamp it out.

I'd been in Africa, of course, without ever knowingly
seeing a malaria-infested mosquito. But here I was in
Maryland, in a strip mall just minutes from where I'd lived
for twenty years, and I'd stumbled across more than 100
million of them, alive and well—and poised to reproduce
virulently.

HOW MANY TECHNICIANS DOES IT
TAKE TO UNSCREW A MOSQUITO?

For my tour of the lab, Hoffman grabbed his white lab coat
off the hook on the back of the door and found a blue one
for me to wear that made me look like a hospital orderly.
We first stopped by the desk of a young woman named
Asha whose job it was to breed parasites. "We create more
parasites in one day than the rest of the world does in an
entire year," Hoffman bragged. He is given to statements
like these, dramatic calculations that do not fail to impress
even though their documentation is unclear.

The mosquitoes were being bred in large beakers that looked like decanters for fine wine. The larvae and pupae moved jerkily in a cloudy yellow liquid. These specimens would eventually be irradiated for about three minutes at a dose that had been determined to yield the greatest protection, carefully balanced to leave the parasite living within weak enough not to do harm but strong enough to trigger the immune system. After that, they'd be taken into another room. There, lab technicians, peering through large microscopes, were dissecting the salivary glands of mosquitoes and extracting the parasites. After that the attenuated, or weakened, parasites would be frozen by a cryopreservation process. In essence, that means they'd be put into a deep freeze, their temperature lowered below zero so that they could eventually be revived and restored to the same state as before, and then stored. "And so those will go into the vaccine?" I asked. "Those *are* the vaccine," Hoffman replied. Once Hoffman completed trials and received FDA approval, he said, the irradiated parasites would constitute the vaccine.

When Nussenzweig had made her key discovery almost forty years ago, she and her research colleagues had said that the trial had "demonstrated for the first time that a pre-erythrocytic vaccine, administered to humans, can result in their complete resistance to malaria infection." The battle against malaria, however, was far from over. As Ruth Nussenzweig and her team put it, "since infected irradiated mosquitoes are unavailable for large scale vaccination, the alternative is to develop subunit vaccines." In other words, vaccines would have to be made from purified pieces of the parasite, rather than whole specimens.[3]

Nussenzweig's approach became known as an "attenuated sporozoite vaccine." Sporozoite literally means animal seed and is the name for the cell form that infects a host, such as the parasite cells that eventually leave the mosquito's salivary gland and enters a human's liver.

The bottom line, as Hoffman explained, "is that it has always been considered clinically and logistically impractical to immunize large numbers of susceptible persons with the irradiated sporozoite vaccine, because the sporozoites must be delivered alive, either by the bite of the infected mosquito, or potentially by intravenous injection, as is done with mice." The challenge, then, is first to make an appropriately irradiated parasite, and then to keep it frisky until it can be injected into the target population. All vaccines can be weakened by time and temperature variations. Many a parasite could perish en route from a laboratory in Maryland to a field station in Africa.

It is in this space between impractical and impossible that Hoffman has decided to bet the ranch.

Sanaria employed twenty-seven people when I first toured the facility. As we made our way through the office, every section was compartmentalized. There were double-door safeguard systems so that one door wouldn't open until the other had been closed, to ensure that nothing could escape. A keypad code was required to enter sensitive areas, and in the chamber between the two doors was the ubiquitous blue-light bug zapper. Protective gear was required, but Hoffman gave me the grand tour. I saw the room where the technicians peered into microscopes to dissect the salivary glands, the room where the mosquitoes were irradiated, and the room where the cultures were bred.

And then as Hoffman speculated about the potential of the vaccine, he got excited again, as he had been on the phone when I first walked in. He told me that one technician working for one hour could dissect a hundred mosquitoes, and that eight technicians working for four hours could produce enough sporozoites to fill the initial clinical trials. Four technicians working for a year could provide enough for the entire military market, and ninety technicians could produce enough for Africa.

He also explained that, to satisfy the FDA, you have to be able to make four guarantees. He summed them up this way, explaining how Sanaria is meeting each requirement:

> First is sterility. Mosquitoes are usually bred in swamps or insectaries. But we are breeding in test tubes and ensuring there is no bacteria or fungi.
>
> Second is purity. We've developed a way of producing aseptic sporozoites and purifying sporozoites that has never been done before.
>
> Third is stability. Can we preserve them in a bottle so that the vaccine will retain its potency when stored? Remember, this is a live, attenuated vaccine, not a dead vaccine.
>
> And finally, safety, that it will not cause malaria in humans.

I asked why everyone had been wrong about how many sporozoites could be extracted. "No one actually bothered to find out, including me," he said. "I was just on the phone yesterday with Ahvie Herskowitz from the Institute for OneWorld Health. He asked me how in the world we kept getting better and better numbers. I said 'Avi, remember the joke about how to get to Carnegie Hall? Practice, practice, practice,'" and at this he threw back his head and laughed.

As we were leaving the lab, I asked Hoffman what could stop him from succeeding. He perked up, as if in anticipation of his own answer: "Nothing! Money, of course, is always an issue. And the security of this lab. If some mosquitoes got out and a man across the street came down with malaria that would be it. I'd be dead. Finished. Just meeting the regulatory requirements to build this place was amazing."

MAN, NOT MYTH

No one person stood out as the obvious and logical choice around which to tell this story. In fact, it was quite the opposite. There were many amazing possibilities from which to choose.

The field of global health is home to Nobel Prize–winning scientists, conquerors of disease, revered humanitarians, and unfathomably wealthy philanthropists. It boasts entrepreneurs like Craig Venter, who won the race to map the human genome, and Victoria Hale, the former Food and Drug Administration official who created the first nonprofit pharmaceutical in order to address neglected diseases. There are leaders of large institutions like the National Institutes of Health, or the Walter Reed Army Hospital, and physicians who have opened small clinics in the most remote jungles and deserts, such as Rick Hodes, who moved to Ethiopia on behalf of the American Jewish Joint Distribution Committee and adopted more than a dozen children in need of complicated surgeries.

But I wasn't searching for the perfect choice. Perfection eludes most of us. Imperfection is more representative. It is certainly more universal.

I sought out Steve Hoffman after becoming intrigued by the way others referred to him, particularly within the tightly knit group of doctors, research scientists, and military and diplomatic officials known as "the malaria community." It was not what they said so much as what they left unsaid. His name invariably left an invisible but palpable tension in the air, like one of those high-energy transmission towers that can be valuable or dangerous, depending upon your point of view.

Without knowing anything else about him, I could sense that Hoffman was a complicated man, someone who challenged others' comfort zones and vigorously protected his own, whose ideas were too radical to simply accept, but grounded in too much experience to casually dismiss.

Fifty-six years old when we first met, trim and muscular, Hoffman had the guarded and intense demeanor of a competitor watching the game clock run down before his victory is secured. He is skilled at political positioning but lacks the politician's gift for small talk aimed at surfacing any patch of common ground that can serve as the basis for a relationship.

He first agreed to see me after receiving a brief e-mail that I'd sent without benefit of introduction from any third party. By coincidence, we'd both graduated from the University of Pennsylvania, and his lab was in my neighborhood. Other than that we had little in common.

Once we met, though, I began to feel some vague but unarticulated kinship with Hoffman, notwithstanding the fact that our personalities were very different. We had both made the transition from long government careers to long-shot start-up enterprises. We'd both worked in institutions— the navy and the U.S. Senate—that afforded resources, prestige, and access to almost anyone or anything one might need. We'd both traded that away for the pressures and headaches, but most of all the freedom, that comes with a start-up enterprise housed in crowded, makeshift offices and financed paycheck to paycheck.

I couldn't walk through his crowded and cluttered lab without thinking of Share Our Strength's first days in the sub-basement of a Capitol Hill townhouse that had been converted from an electroshock therapist's facility, complete with sound-muffling egg cartons glued to the walls. I remembered that feeling of having the kernel of a half-baked idea that the rest of the world had yet to hear about or understand, but that, once developed, tested, and refined, might prove to inspire and mobilize others.

I certainly didn't put Hoffman's odyssey at the center of this story because I had the foresight to be sure he would succeed. Indeed, the odds of him reaching his goal are long, if not forbidding. We won't know the full measure of Hoffman's success or failure, or that of any of his competitors, until the passage of time has had its way. Many years will be required for conducting and assessing clinical trials. Even if his vaccine makes it through those hurdles and a successful vaccine is put into wide use, the malaria parasite could evolve to escape defeat, as it always has in the past. There are an infinite number of variables, ranging from climate change to African infrastructure, that may have more to do with whether the vaccine works on the ground than anything Hoffman does or doesn't do in the lab. And there are other possible breakthroughs on the horizon that could blow Hoffman's ideas out of the water. Scientific discovery, by its very nature, stands still for no man.

But the trajectory of Hoffman's life and career so clearly parallels and illuminates our society's changing approach to solving social problems. He began as a doctor doing what doctors do, helping one person at a time. But as he became interested in tropical diseases like dengue fever and malaria, he came to see that the scale of the problem and the enormous number of people affected was far greater than what any one doctor could handle. It was greater even than what all the doctors in the field of tropical medicine could handle. And the problems were not just medical, they were economic and political.

When he realized it would take the resources of government to solve the problems he cared about on the massive scale on which they existed, he joined the U.S. Navy, which had the best facilities at the time, and eventually led its malaria vaccine development efforts. The goal was not just delivering good medical care, but scaling up that care so that others would have access to it. Lacking an economic market for doing so, Hoffman found a political market in the form of government. For twenty-one years, the U.S. Navy and Army offered the tools necessary to advance vaccine development.

But after a certain point, he also came to see the limitations of what could be done via government. He then became the classic entrepreneur, resigning from government, setting out into the private sector, and starting a company— a biotech company. He chose to operate at a new intersection of philanthropy and entrepreneurship that would permit him to take risks and try out innovative ideas in order to solve problems that there were no economic or political markets for solving.

THE "SPACE RACE" OF
THE TWENTY-FIRST CENTURY

The accelerated and massive investment in global health and in the eradication of diseases affecting the poorest people on the planet has been a powerful generator of ideas and strategies in the field of health care. But, like the space race of the twentieth century, it has applications that reach beyond its own immediate field to impact other social challenges. Outside of government, the work of global health is conducted through nonprofit organizations. Generations of social-change agents in every field will be shaped by what is happening in global health today. And the catalyst for investing in global health has been the Bill and Melinda Gates Foundation. It is the modern day NASA of the global health field.

In the 1960s and 1970s, NASA-led space programs, from Mercury through Apollo, yielded thousands of spin-offs, adaptations, and alternative uses that have impacted every aspect of life. They range from kidney dialysis machines that were derived from processes to remove toxic waste to smoke detectors first used in Skylab that are now common in almost every home, from the fabrics of fire fighters' uniforms to ear thermometers, from solar energy panels to weather forecasting and water treatment systems for developing nations. Few Americans have a direct connection to the men and women who have gone into space or the team that supported them. But no American was left untouched by the literally thousands of applications of the technologies created for space. The goals of the space race pushed the edge of the envelope of innovation and inspired some of the best minds of a generation to achieve things that reached far beyond the parameters of the space race itself.

President Kennedy was able to foresee the impact of the space race when he announced the challenge at Rice University on September 13, 1962. What he said can be instructive as we face a new kind of challenge—a challenge to improve life on earth:

> We choose to go to the moon in this decade and do the other things, not because they are easy, but because they are hard, because that goal will serve to organize and measure the best of our energies and skills. . . .
>
> The growth of our science and education will be enriched by new knowledge of our universe and environment, by new techniques of learning and mapping and observation, by new tools and computers for industry, medicine, the home as well as the school. Technical institutions, such as Rice, will reap the harvest of these gains.[4]

Thousands of scientists, researchers, manufacturers, computer programmers and contractors from all around the world became part of the NASA effort in the same way that the goals set by the Gates Foundation have mobilized thousands of doctors, scientists, biotech companies, labs, and universities, bringing new talent into the effort at an unprecedented pace. Global health spending will result in new medicines, vaccines, cures, and treatments for diseases and health-care practices. It has already produced new diagnostic techniques, new kinds of sterilization and purification equipment, new preservation methodologies, and an entirely new field, that of synthetic biology.

But it is also leading to the creation of innovative new financing mechanisms, such as advanced market commitments and philanthropic collaborations between governments and foundations, and has produced the first nonprofit pharmaceutical. It is even leading to new, clean energy sources.

As the modern-day equivalent of the space race, our global health challenges will transform the nonprofit and philanthropic universe in ways far greater than anything we might have anticipated, changing the way we approach a vast number of social problems.

In June 2004, Hoffman applied for one of the grants that the Bill and Melinda Gates Foundation was making, especially designed for risk-taking projects aimed at making big breakthroughs—like solving the malaria vaccine issue. He did not get it.

CHAPTER 5

TROPICAL LINEAGE

A team of researchers in Costa Rica's Alberto Manuel Brenes Reserve have been searching for plants that might help cure the mosquito-transmitted disease known as malaria. . . .

During their research, the team collected a total of 50 promising plants. . . . As of now, no other details have been released by the team as to why they think that these species . . . might cure (or help prevent?) malaria.

—Levi Novey, "Potential Cure for Malaria
Discovered in Rainforests of Costa Rica,"
EcoLocalizer, September 18, 2008

TROPICAL REGIONS ARE PARTICULARLY FERTILE areas for disease because of the many insects that breed and thrive there. Tropical medicine is the most dangerous and least lucrative of all medical practices. Patients live far away in jungles and near swamps. They are poorly educated, have little money for medical bills, no insurance, and barely enough food or water to survive. They are exposed to horrific diseases that often have no cure, and of which most of us have never heard.

One of them is African trypanosomiasis, also known as sleeping sickness, which infects half a million people annually and kills 50,000. Visceral leishmaniasis, transmitted by sandflies, afflicts 1.5 million every year, giving them skin ulcers and potentially massive tissue destruction, and is 90 percent fatal if untreated. Chagas' disease infects 16 million to 18 million people annually, mostly in Latin America. The parasite enters through broken skin, and there are often no symptoms for years, yet the disease invades organs and creates severe cardiac problems, claiming some 45,000 lives a year.

A magnified photo of just one of the parasites that teem within human hosts would deter most Americans from ever applying for a passport.

THE PRIESTHOOD OF TROPICAL MEDICINE

As Donald Burke, past president of the American Society for Tropical Medicine and Hygiene, explained at the centennial meeting of the organization in 2003:

> It is no coincidence that our society was founded at the very moment when the USA first emerged as a global power. . . .
>
> After the Spanish American War of 1898, the United States suddenly found itself with a string of new possessions that almost circled the globe in the tropics, including Cuba, Puerto Rico, Hawaii, the Philippines, and various

island territories in the Pacific Ocean. U.S. military personnel sent to occupy these new tropical possessions were decimated by infectious diseases.[1]

Doctors who specialize in tropical medicine went through the same rigors of medical training and accumulated the same amount of medical-school debt as their colleagues who chose pediatrics, oncology, cardiology, or neurology. But when they chose different diseases, they chose different patients. As a result, they gave up the comforts of American medicine and many of its rewards.

The diseases that plague much of the world are neglected across the entire spectrum of American health care, from medical schools to research hospitals to global pharmaceuticals. Doctors Without Borders reports that of 1,393 new medications introduced into the market between 1975 and 1999, only 13 (or about 1 percent) were for the treatment of tropical diseases.[2]

Only 10 percent of global health research is devoted to diseases that account for 90 percent of the global disease burden. Meanwhile, billions of dollars are spent each year on research and development of drugs for ailments that afflict people who can pay, which include conditions such as obesity, baldness, and those associated with aging among Baby Boomer populations.[3]

In 2002, Michelle Barry, another past president of the ASTMH, explained one reason why we should be concerned: "In the medical condition of the world's poorest people we can see the incubators of political and social pathology as well as medical, and as events of the past year have pointed out, the borders of the advanced industrial countries are permeable to all three. Tropical medicine specialists are a kind of distant-early-warning system of public health. We see problems in their early stages."[4] In other words, in a world where global travel is common, those diseases are no longer so far away.

Tropical medicine is a priesthood of sorts. Passionate true believers prize purity over profit, sometimes even over success. They may be among the last of the idealists. If there is a way to practice medicine and not make money, this is it.

So, when a renowned leader in tropical disease like Steve Hoffman transitions from a lengthy career in government service to launch his own commercial business venture, it can feel, to some, almost like a betrayal. Like going over to the other side. Repudiating the path others have chosen and clung to. Selling out. For-profits are suspect. The pressure of having to make a profit has always prevented pharmaceutical companies and health-care corporations from investing in R&D and clinical trials for drugs unlikely to find a profitable commercial market.

For-profit or not-for-profit, the odds are stacked against anyone who has enlisted in the battle against malaria. Historically malaria has proven the most difficult to conquer of all tropical diseases. The parasite transmitted by certain species of mosquitoes has proven to be remarkably resilient, mutating, adapting, and eventually growing resistant to every drug that has been designed to attack it. Hoffman himself, describing the challenge of stopping malaria, led one reporter to conclude, "It's always a good bet to put your money on the parasite."[5]

Those who persist in developing malaria drugs and vaccines are like increasingly desperate homicide detectives on the cold-case squad, obsessed by a serial killer who has managed for years to evade, elude, and outwit them at every turn. Having pursued all the obvious leads and logical clues, they are unwilling to admit defeat, though all that remains are ever more fantastic theories and complicated scenarios. Still stymied after exhausting all of the practical ideas, what's left are only the impractical ones.

Sometimes, though, progress creeps along an inch or so at a time. Like good detective work, good science often comes slowly and in small steps after years of persistent, methodical legwork. Chasing down leads, eliminating hypotheses, piecing together facts, and applying instinct and judgment can take time. Good science requires patience, because it requires waiting: for tests, trials, reviews, corroborations, and approvals, and for insight, understanding, and synthesis. For children exposed to malaria, waiting often means dying. Every thirty seconds, a child under five years old dies from malaria, as the clock ticks off months and years. Scientists like Steve Hoffman race each other and race against time.

Malaria conjures a distant place and another era. One thinks of stifling heat and humidity, fetid swamps, Africa's thatched-roof huts, nineteenth-century explorers, khakis, medical tents, ceiling fans, bed nets, and, of course, mosquitoes. It is not only an ocean or a century that separates us from malaria. So does air conditioning, urban density, building codes, technology, sophisticated health care, and education. Malaria is about Africa, and as Americans we don't go much to Africa. We don't live, study, or travel there. Americans only rarely contract malaria or die from it. Most of us don't know anyone who ever has.

The malaria parasite is one of the oldest and most persistent of all diseases, and throughout the course of human history, the battle against it has been a losing one, each advance met and outwitted by the parasite's ability to mutate and resist drug treatment. No more epic struggle between life and death has ever been waged on planet earth.

It's not surprising that Steve Hoffman, with his sterling pedigree and a track record of having made good use of it, chose the fight against malaria as his fight.

THE "GRAND OLD DAYS"
OF TROPICAL MEDICINE

Hoffman was born and raised in a middle-class family near Asbury Park, New Jersey. His father sold paper products. He majored in political science at the University of Pennsylvania. He took off his junior year, traveled abroad, and on a visit to Jerusalem decided that he wanted to dedicate his life to helping people. He called his parents and said he was going to become a doctor. After graduating from Penn he ended up at Cornell Medical School.

At the time, Cornell was one of only two medical schools in the country that made tropical medicine a requirement. Hoffman took the sixty-hour course in his second year of med school. At Cornell he heard lectures by the legendary Dr. Ben Kean, a renowned teacher, researcher, and practitioner of tropical medicine. (In 1979, Kean went on to wider renown as the doctor who brought the shah of Iran to the United States for medical treatment, indirectly setting off the Iranian hostage crisis and fatally complicating the presidency of Jimmy Carter.) Hoffman had the opportunity to join with other students who would gather round Kean and listen to his tales after hours. He was quick to fall under his spell. Kean believed that early hands-on experience in the tropics was the best way to stimulate careers in the specialty. He was instrumental in helping many medical students obtain these experiences. Hoffman applied for a fellowship to spend the next summer studying lactase deficiency in infants in Bogotá, Colombia. He ended up staying in South America for a year. When he returned, he was committed to a career in tropical medicine.

In his training and commitment Hoffman descends from a distinguished lineage that covers almost the entire history of modern tropical medicine. Kean's mentor was Francis O'Connor, his professor of tropical medicine at Columbia University's College of Physicians and Surgeons in 1934. O'Connor was an unforgettable figure. In his autobiography, Kean describes him striding into the lecture hall:

It was a breathtaking entrance. As was his custom, O'Connor wore morning clothes to class—grey pinstriped trousers and a cutaway coat. He carried a bowler hat and an ebony cane. A hundred second-year medical students swiveled from their microscopes and gaped in wonder. O'Connor, elegant, imperious, civilized, stared back, as if he were peering through his microscope at a wholly predictable, if mildly amusing species of protozoa. . . .

And how he could teach! O'Connor drew crayon sketches for me in his quick, deft hand of all the important parasites and their complicated life cycles—"mug shots" he called them. Starting with malaria, he explained how all parasitic diseases have three things in common: they all feature a scene of the crime, the human body or the host, where the evil business is done; each involves a third-party accomplice called a vector that helps transmit the infection (in malaria, the mosquito); and all involve a cunning culprit, the parasite itself.[6]

When Kean told stories to his students, he was following in his mentor's footsteps. O'Connor had spent hours regaling Kean and his classmates about the "grand old days" of tropical medicine. The hero of his tales was the widely acknowledged father of tropical medicine, the founder in 1899 of the famed London School of Hygiene and Tropical Medicine: Sir Patrick Manson, also known as "Mosquito Manson." It was Manson who had discovered, in China, that the mosquito was the transmitter of filarial, the parasite which causes elephantiasis, and who later suggested, but never proved, that it might also be the transmitter of malaria as well. And it was Manson who mentored Ronald Ross, a British officer in the Indian Medical Service who was the first to conclusively demonstrate that malaria parasites were transmitted by mosquitoes.

It was a French army surgeon, however, who was the first to notice parasites in the blood of a patient with malaria. Alphonse Laveran, born in Paris in 1845, had decided to follow in the footsteps of his father and become a military doctor. In 1878 he was posted to a military hospital in what was then the French territory of Algeria. While working there, Laveran saw dark, pigmented granules in the blood he examined, and motile, flagellated bodies.

Followers of Louis Pasteur had already suspected that "bad air" was not to blame. (The conjecture previously was that malaria was caused by mysterious "vapors," mostly from swamps, and the name of the disease, from the Italian *mala aria*, literally means "bad air.") Pasteur was convinced of a bacterial cause until Laveran wisely invited him to look through his microscope at what Laveran described as "filiform elements which move with great vivacity."[7] Pasteur was quickly convinced.

Laveran was prolific. He wrote six hundred scientific papers and six books, and in 1907, his discovery was recognized with a Nobel Prize.

Identifying the parasite solved one mystery, but for the next twenty years it remained unclear how the parasite found its way into humans. In 1897, Ross demonstrated that malaria parasites could be transmitted from infected patients to mosquitoes and then back to other humans who would become infected in turn. It was in August of that year, while dissecting the stomach tissue of an Anopheline mosquito, that Ross found parasites and proved the role of mosquitoes in transmission.

Ross became an expert at dissecting mosquitoes, something that Steve Hoffman would eventually excel at as well, and his painstaking research took years to bear fruit. But when it did, the discovery that mosquitoes were the vector by which malaria was spread won Ross a 1902 Nobel Prize.

In his December 12 lecture upon accepting the award, he placed the issue in broader context:

> Malarial fever is important, not only because of the misery which it inflicts on mankind, but because of the serious opposition which it has always given to the march of civilization in the tropics. . . . There it strikes down, not only the indigenous barbaric population, but, with still greater certainty, the pioneers of civilization, the planter, the trader, the missionary, the soldier. It is therefore the principal and gigantic ally of barbarism. No wild deserts, no savage races, no geographical difficulties have proved so inimical to civilization as this disease. We may almost say that it has held an entire continent from humanity.[8]

The same words could be spoken today with the same veracity.

Dr. Ben Kean went on to work at Gorgas Hospital in Panama, which he described as "the mecca of tropical medicine." It is named for William Crawford Gorgas, who helped to eliminate yellow fever and reduce malaria during the building of the Panama Canal. Of the 26,000 men working on the Panama Canal in 1906, 21,000 were hospitalized for malaria at some point. Hoffman's training would one day include numerous visits to Gorgas as well. And a trip to Ecuador while on break from medical school in December 1972. He ended up hospitalized and alone for ten days with typhoid fever, and the experience cemented his commitment to tropical medicine.

Hoffman did his residency in family medicine at the University of California–San Diego and practiced there from 1975 to 1978. He spent 1979–1980 establishing the Tropical Medicine and Travelers Clinic there.

But practicing tropical medicine in the United States made him feel "like a fake tropical medicine doc," he told me. "So after all of the years of antiwar protest and all the rest, I cut my hair and joined the navy so that I could go to their research station in Indonesia and study malaria." In April 1980 he went to what is known as NAMRU-2, the Naval Medical Research Unit Two, an infectious disease research facility established in Guam in 1943 by the Rockefeller Foundation and eventually relocated to Jakarta, Indonesia. Its primary function has always been to study infectious diseases of military significance because they might threaten mission readiness.

Within forty-eight hours of his arrival in Jakarta, Hoffman was treating patients with typhoid. In ten weeks he saw seventy such patients, and 18 percent of them died. As he told the American Society of Tropical Medicine and Hygiene, "I had been taught . . . that the mortality rate for Typhoid had been dropped to less than 1%, and I was appropriately horrified, wondering what we were doing wrong. It was clear that there was a major disconnect between what I had been taught about treatment of typhoid, what I as a well-trained physician could do for a patient with severe typhoid, and the outcomes for hundreds of patients in that hospital."[9]

Hoffman spent five years there, from 1980 through 1984. While working on typhoid fever, malaria, cholera, and dengue fever, Hoffman had the experience of every doctor who has practiced there: "countless children died in my arms." He and his colleagues experimented with different approaches until settling on the use of a high dose of a drug called dexamethasone. In 1984 they published a study about it in the *New England Journal of Medicine*, and it became the new standard for treating severe typhoid. By the time he left Jakarta, the fatality rate for typhoid had dropped to 1 percent.

After that he came back to the United States to join the navy's malaria vaccine development unit in the Maryland suburbs of Washington, D.C., and ultimately to head the malaria program. He also found a way to stay close to patients who could not afford health care, working in the emergency room of Providence Hospital, a community hospital chartered in 1841 by President Abraham Lincoln in what is now one of Washington's poorest neighborhoods.

Hoffman's naval career totaled twenty-one years. Having initiated the Department of Defense's Plasmodium genome sequencing effort in 1995, he ultimately left to work at Celera. Meanwhile, he became president of the American Society for Tropical Medicine. It was during this time that he wrote most of his 350 scientific papers, while also obtaining numerous patents.

Along the way he won every professional distinction possible, including the Bailey K. Ashford Medal from the American Society of Tropical Medicine and Hygiene in 1992; the Legion of Merit from the U.S. Navy in 1993 and 2000; the Col. George W. Hunter III Certificate in 1994; and the Captain Robert Dexter Conrad Award, the U.S. Navy's most prestigious award for scientific achievement by anyone in any field, regardless of military affiliation, in 1998.

Mapping the malaria genome was enormously complex, taking $28 million and a lot longer than Hoffman or anyone else anticipated. In fact, it took an international consortium of organizations with scientists from more than a dozen institutions. But the genome work, of which many other experts were at first skeptical, had incredible influence on the field and changed the way malaria research is done. It provided the opportunity to find the genes from the *Plasmodium falciparum* parasite that would make the best vaccine and drug targets: a chance, as one researcher put it, to "search for the chink in the parasite's armor."[10]

It also had an important influence on Hoffman. "I realized from all of those different lines of work, all of which people said couldn't be done, that if you actually organize yourself well, get the right smart people around you, put the story together properly, you can actually accomplish a lot of things everybody said were impossible or impractical," he told me.

In the spring of 2002, Hoffman organized a scientific summit through the Keystone Symposia, an independent nonprofit devoted to advancing biomedical and life sciences. One session was titled "Malaria Vaccines: Why Is It Taking So Long?" Hoffman was struck by the predictions of various experts as to when they thought a malaria vaccine might be launched as a commercial product. "Many in the room thought it would be sometime between 2020 and 2025 at the least, even with the genome," he said.

Hoffman thought back to the approach that was pioneered by Ruth Nussenzweig at NYU when she showed that she could protect mice against mouse malaria. In the early 1970s, there were two separate groups of researchers who had thought that the same strategy would work on humans and had seemingly proved their point. "It was shown independently that you could protect people by this method, but it appeared to be totally impractical," Hoffman said. "But this is the lesson I learned: like sequencing the malaria genome, this was a bioengineering problem, not a scientific discovery problem."

With this one insight, Hoffman had fundamentally reimagined and redefined what it would take to create a vaccine to prevent and ultimately end malaria. It was as if he had looked down on the playing field and realized that it would not be good enough to be the first across the goal line; rather, the vaccine development community was playing on the wrong field. Success would require mastering not just the traditional disciplines of biology and immunology, but those of engineering and entrepreneurship as well. It wasn't that world-class scientific discovery would not be necessary. After all, Hoffman would go on to assemble a team of accomplished scientists from around the world. It's just that science alone wouldn't be sufficient.

When Hoffman described this "bioengineering" challenge to me, I was struck by the similar problem faced in so many dissimilar fields, albeit fields with which I had greater familiarity. In the social sciences it is often presumed that because of the persistence of seemingly intractable problems—hunger, homelessness, teenage pregnancy, unemployment, drug abuse, and so on—we don't know what the solutions are. In almost every case there are programs that work, programs that address such problems with high rates of effectiveness. But they tend to be local, idiosyncratic, and impractical, for one reason or another, as models that could be used elsewhere. Just as in the case of the Nussenzweig vaccine that Hoffman hopes to manufacture, the challenge is not one of discovering new solutions, but of making the solutions that have already been discovered affordable, replicable, scalable, and sustainable.

CHAPTER 6

BATTLEFIELD GENERAL

Researchers in Melbourne believe their discovery could be a major break-through in the fight against the disease.

The malaria parasite produces a glue-like substance which makes the cells it infects sticky, so they cannot be flushed through the body.

The researchers have shown removing a protein responsible for the glue can destroy its stickiness, and undermine the parasite's defence. . . .

Professor Alan Cowman, a member of the research team at the Walter and Eliza Hall Institute of Medical Research, said targeting the protein with drugs—or possibly a vaccine—could be key to fighting malaria.

"If we block the stickiness we essentially block the virulence or the ca-pacity of the parasite to cause disease," he said.

—Phil Mercer, "'Breakthrough' in Malaria Fight," BBC, July 14, 2008

ON MY SECOND VISIT TO Stephen Hoffman at his Sanaria office and lab in Rockville, he was waiting for me, and his assistant ushered me into his office immediately. I had the sense that having someone write about him was appeal-ing to Hoffman, that he viewed it as external affirmation that his journey, scoffed at in some circles, was worth observing. In the environment he works in, which is intensely competitive and also marked by routine failure and only occasional breakthroughs, such affirmation is rare. You take it where you find it.

But he was also cautious, wary of the idea of an outsider writing about his work, and wary of me, whom he didn't really know. Much of the lab's methodology is proprietary, and Hoffman did not want to see it compromised. The methodology is not only critical to the development of a malaria vaccine but could have applications to many other biologic and scientific processes. Sanaria has had to learn how to harvest record levels of parasites from mosquitoes, separate them from other potentially infectious salivary gland material, store and preserve them—all in ways that have created valuable new intellectual property. Hoffman is nothing if not a competitor. And he knows he's in a marathon race. He's busy, focused, and disciplined.

We talked about how his work was different from what he did in the navy for twenty years. As he spoke, he struck me as someone who was ready to take on all challengers. Hoffman has his supporters, some of whom are impressive overachievers in their own right, such as the late Maurice Hilleman, who, at Merck, developed eight of the fourteen vaccines now given to children routinely, and who was one of the first to join Sanaria's board. But there's an intensity to Hoffman, the wariness of a man who knows he has skeptics. His chin tilts up a bit, and the air of impatience rarely leaves him.

Casual in black jeans and cowboy boots, he was fit and trim, even a bit tan. Surrounded by wood-carved art from New Guinea and pictures of his wife, Kim Lee, he tried to explain how he saw his role now compared to his navy years:

> The way I think about my work is not much different. I never really felt like I had a boss in the navy. I could do what I wanted to do. But this is bioengineering, not scientific discovery. Our job is to develop and manufacture a vaccine. That's it. That's what we have to stay exclusively focused on. There are plenty of colleagues who are smarter than I am. What I'm good at is having a vision of what is possible, and putting together all of the pieces to achieve it. Someone has got to be the one to say this is achievable. Most people aren't able to envision something that hasn't been in their line of sight before. And so my job is to keep everyone focused on producing the vaccine, that's it, that is all we are about. And I just keep driving them to do that.

His ambition was for production at lightning speed. "We'll need 3,000 doses for toxicology trials," he told me. "After that it goes to the FDA. I'm not expecting any surprises. They can turn around an answer in thirty days. I'm twelve months away from putting this vaccine in an arm, and eighteen months away from putting it in babies in Africa."

As confident as he was, Hoffman was quick to interject a note of caution. "But who knows? A lot of people have tried and failed before. I've tried and failed. The first thing I ever did was develop a treatment for severe typhoid fever. That was in 1984. Other than when I was an emergency room doc, there's nothing I've done since that has saved a single human life."

That success is the exception rather than the rule seems to be the accepted dynamic in laboratory science. It is the nature of science, the logarithm of scientific advancement and achievement. Most scientists acknowledge it no matter how begrudgingly they accept it. Still, it is one thing to say it and another to live it. Living it means logging late nights and long hours in the lab, measuring and marking and checking and double-checking tests and trials cobbled together from grants that took weeks to write and then months to hear whether they were approved or rejected. Living it means writing highly technical papers that must be peer-reviewed and that, when published, will draw criticism from others with only half the experience.

In addition to a little bit of luck, it takes a special personality to pull off something like the development of a vaccine: It takes someone with persistence bordering on stubbornness, confidence bordering on arrogance, the long-term patience of a cathedral builder, and the immediate impulses of an emergency-room doc. It takes leadership. And it takes a boxer's willingness to take a punch and come up off the canvas.

I was curious how Hoffman's vaccine was different from the better-known, better-funded GlaxoSmithKline (GSK) vaccine candidate, RTS,S, which was farther along in clinical trials. In 2005, clinical trials in Mozambique for the RTS,S reported efficacy of 29.9 percent, and at the end of the six-month observation period, prevalence of the *P. falciparum* parasite was 37 percent lower in the RTS,S group compared with the control group. Later trials would show the effectiveness of RTS,S jumping up to 53 percent, an impressive gain, but one that still would leave nearly half the population unprotected.[1] Such progress at least ensures continued funding. More than $800 million will be invested in RTS,S before all is said and done, an amount that dwarfs the resources available to Steve Hoffman. If there was a front-running candidate among the various malaria vaccines under development, RTS,S was it.

RTS,S and Hoffman's vaccine are both built on what is known about the circumsporozoite protein that the Nussenzweigs identified as a vaccine target. Hoffman was part of the team composed of the Naval Medical Research Institute, Walter Reed Army Institute of Research, and GSK that ran the first clinical trial of such a vaccine. GSK "kept going, and I diverged," Hoffman explained in an interview with the tropical disease website TropIKA.net. "As Director of the Navy malaria program, my job was to develop a vaccine that could be used for military personnel or travellers, meaning that it had to be at least 80%, preferably over 90%, protective to have the operating characteristics of most other vaccines. . . . I've always felt that's the kind of vaccine we need for everybody—that there shouldn't be different tiers of vaccines: one for travelers and another for kids in Africa."[2]

At one point in our discussion, Hoffman became animated and went into to another room to get the large laminated malaria life-cycle chart that is the indispensable teaching tool of all malariologists. The illustration is the anatomical outline of a human body from about the shoulder to mid-thigh. It shows a mosquito on the arm and then the rapid flow of parasites through the body, first to the liver, and then bursting out of the liver in a new, more mature cellular form. Now "merozoites," they head into the red blood cells to do their greatest damage. Hoffman laid the chart down on a table and demonstrated strategy like a four-star general explaining a battlefield map. He leaned over it, pointing out critical areas with broad sweeps of his hand.

Hoffman used terms like "merozoite invasion" and spoke of the parasites' skill at "evading defenses." He explained the possible options and strategic choice he had made, dividing the diagram into three sections, surveying the various battlegrounds: the transmission stage, which represents the other part of the parasites' life cycle as it gets into and out of the mosquito; the liver stage, or pre-erythrocytic stage, meaning before the merozoites get to the red blood cells; and the final stage where the red blood cells are being affected.

Hoffman said he is not that interested in campaigns against the merozoites entering the red blood cells. Nor is he interested in the air war of a transmission vaccine—which is also known as an "altruistic vaccine," because it doesn't help someone already infected but instead protects the rest of the community by blocking transmission of the parasite from one mosquito to another.

Hoffman has chosen instead to target the liver as his Omaha Beach. "This is where we've got to stop them: in the pre-erythrocytic stage. If they can't get out of the liver, they die. The T cells will kill them," he said, speaking of a natural ally the way General Dwight D. Eisenhower during World War II might have spoken of the Brits.

Years of laboratory reconnaissance have revealed quite clearly how the enemy operates. There are two places where the parasite does its work: One is inside the mosquito, the other is inside human beings. Both represent opportunities to put the enemy out of commission.

JOURNEY TO THE
CENTER OF THE MOSQUITO

When a mosquito bites an infected person, it ingests the parasite at an early developmental stage. Once inside the mosquito's midgut the parasites will develop into what are called "sporozoites," some of which make their way to the mosquito's salivary gland. The next time the mosquito bites someone, hundreds of sporozoites will be injected into the new victim.

Here's how the mosquito pulls it off: A female Anopheles mosquito, hungry for blood, lands on a patch of human skin. Only the female mosquitoes bite, and it's only because they need the nutrients and protein of a blood meal to be able to make and lay their eggs. Female mosquitoes mate only once, but they store enough sperm to use throughout their reproductive life.

When it bites, the mosquito probes with a long, needle-thin, tube-like proboscis that actually has four tools inside of it: Two have serrated edges to slice and drill a hole in the donor's skin, one acts like a hose to inject saliva, and the other is like a straw to draw blood into the mosquito's body.

The authors of an article in the *Johns Hopkins Public Health* magazine described the mosquito's actions in vivid terms:

At the end of the proboscis, knife-like stylets move rapidly like electric carving knives to split the skin. She gently jabs at different angles in the hole until she nicks an arteriole that spouts a subcutaneous pool of blood that she can draw from. Exquisitely evolved, the female vampire will squirt into the cut a small amount of saliva full of anticoagulants to prevent the blood from clotting.

Within a couple of minutes, her translucent belly bloats and shifts from waxy gray to cherry red. She sucks a few micrograms of blood—more than her own body weight. . . .

. . . And she sucked up something else as well: some protozoan stowaways.

The mosquito, in a simple act essential for reproduction, ensures the reproduction and spread of another species: the *Plasmodium* parasite.

The malaria cycle begins once more.[3]

Compared to the toll taken on human beings, the toll on the mosquito is minuscule: The mosquito herself gets off quite easy, an unwitting carrier of destruction. It is inside of our bodies that the real carnage takes place.

A mosquito transfers about 10 percent of the parasites it is carrying when it takes that next bite. These sporozoites travel through the blood and reach the liver in less than thirty minutes. In the liver, all hell breaks loose.

The sporozoites, safe in the liver, form "schizonts," large, multinucleated cells that divide and multiply. When the schizonts burst, they release as many as 40,000 merozoites. If there was a factory anywhere in the world that consistently produced product at this rate it would be the globe's dominant brand, which is exactly what malaria is in Africa: the dominant brand-name disease.

By the seventh or eighth day, the liver cells release the merozoites, which head for red blood cells to finish their deadly mission. The merozoites trigger a reaction from the B cells of the host's immune system. Many of the merozoites are destroyed, but those that escape invade the red blood cells. It is when the merozoites burst out of the liver and head into red blood cells that the host spikes a high fever. The parasite forms knobs on the exterior of a red blood cell, which allows it to adhere to cells lining the blood vessel and ultimately impede blood flow.

In her book *Diseases and Human Evolution*, paleopathologist Ethne Barnes explained that "the disk-shaped red blood cells are responsible for transporting life-giving oxygen from the lungs throughout the body tissues and taking away the gas waste product of cellular respiration, carbon dioxide, to the lungs to be expelled. . . . As the number of merozoites increases, the number of viable, circulating red blood cells decreases, producing anemia in the host."[4]

The red blood cells themselves soon burst, releasing more merozoites, which invade fresh blood cells, and the cycle continues, over and over, until billions swarm in the blood.

The parasites replicate in the blood every forty-eight hours, and each one has the chance for mutation. As Carole Long, chief of malaria research at the National Institute of Allergies and Infectious Disease, said, "the parasite you put in is not necessarily the parasite you get out."[5]

WHERE THE WILD THINGS ARE

Hoffman's work isn't the only exciting development in malaria, just the boldest and most imaginative. In fact the field is alive with laboratory research, clinical trials, field studies, and conferences. The World Health Organization is now tracking thirty-five separate vaccine-development efforts. One of these is with a former Hoffman collaborator at the Walter Reed Army Institute of Research.

His name is David Lanar, and I decided to visit him. After all, Walter Reed's research facility was in Silver Spring, only a hop, skip, and jump away from Hoffman's strip-mall offices. As I arrived for my appointment, I noticed a sign just past the institute's main gate announcing that I was in a restricted area and that federal law prohibited me from taking notes on any of the activities therein without the express permission of "the Commander." Security at the various checkpoints was relaxed but thorough. Pairs of soldiers in camouflage fatigues passed by as I waited for Lanar, who was coming up on his twentieth year in immunology at Walter Reed. Despite the posted warnings, he seemed more than happy for me to write down every word he said.

When Lanar arrived in the lobby, he looked like Maurice Sendak's version of a parasitologist, something right out of *Where the Wild Things Are*. He was a heavy-set, round man made up of equally heavy and round constituent parts. He had a large and friendly face, with only his eyes and his ruddy cheeks visible through a crown of black and gray hair and a beard. He had massive hands, and a white button-down shirt strained over his barrel-chest. It was open at the neck, at the sternum actually, and rumpled all the way down to where it was ambivalently tucked into his faded black Levi's. The grey hairs of his chest sprouted through his collar as though reaching up to unite with his scruffy beard.

His pride in the building and its mission was quickly evident. Walter Reed is a massive facility that employs 1,200 researchers, many of them civilian, and almost 400 of the 1,200 are working on malaria. "It is the largest number of malaria researchers anywhere in the world," Lanar explained. "This is also where they bring suspected anthrax for testing." He pointed out a vaccine manufacturing facility across the street that enabled Walter Reed to create vaccines without depending on the large drug companies.

Trained at the London School of Hygiene and Tropical Medicine, Lanar had been at the National Institute of Health, working on other parasites—Leishmania and Shistosomiasis—but left to join the army because of its labs and capacity. Like many tropical medicine docs, he is cast against type for the U.S. military, but there's not much of a market for private practice in this field. Those who want to work in it and want to have the tools necessary for success learn to salute and join one of the largest bureaucracies in the world.

When we got to his office, which was crowded with books and files, I asked why the interest in global disease and malaria had been increasing recently. He and his office mate, Ann Stewart, seemed aware of the revived interest in malaria but only vaguely, as if developments in the world beyond their microscopes were merely distant, rumored events. Stewart, who had a stuffed toy monkey draped over her microscope, attributed some of the increased attention in global health and neglected diseases to the Gates Foundation and the Harvard School of Public Health. She described a compelling dramatization of the continuing imbalance in investment that was presented by Amir Atarran, a lawyer and immunologist who writes and lectures on global health.

Attaran would take the stage with a large jar and a supply of small, round BB's. He would drop in the few that he said represented the world's investment in neglected diseases like malaria, and then he would pour the amount that he said represented the world's investment in HIV. Apparently he would stand there pouring for quite some time, with the racket getting louder and louder, until the point could not have been missed.

Lanar's focus, like Steve Hoffman's, is the development of a malaria vaccine, and for the same reason: Malaria has become resistant to almost all of the drugs that have been developed to fight it, "but we've never had a malaria vaccine so we don't know how it will react." The research process with new vaccines usually goes from table top to animal to human. But "there is no way to conduct animal experiments" with malaria vaccines, Lanar said, "because falciparum malaria is unique to humans. Obviously the FDA has tremendous confidence in our approach, otherwise they would never let us challenge human beings."

Lanar and Stewart described what is known as "the hotel phase" of clinical trials. Volunteers that have been vaccinated are "challenged" with malaria by being subjected to mosquitoes until they are bitten a sufficient number of times. These aren't just any mosquitoes, though; they are carefully chosen ones that are in a small box that is placed over a volunteer's arm—five to a box, each carrying the parasite. The volunteers then check into a hotel with physicians who examine them twice a day for about two weeks. If a significant percentage of those who have been vaccinated resist the disease, the experiment is deemed a success. Those who show symptoms are treated immediately and effectively.

Lanar was cautiously optimistic about the vaccine he had been developing. Known as LSA-1, the liver stage antigen, it attacks the parasite at a mature stage of its development. But he was quick to suppress expectations. Lanar told me of the five years he had spent building the vaccine and said, "I was convinced it was going to make the cover of *Science* magazine. But the vaccine failed, and I have to tell you I got really depressed. I was clinically depressed for quite a while."

It's not surprising. Parasitologists tend to be obsessively committed to their task. Lanar had a stained-glass piece depicting the Trypanosoma parasite, which is transmitted by a tsetse fly and causes Chagas' disease, hanging in his office window. It showed the parasite defecating and invading the heart muscle, all in glorious sun-streaked colors. Lanar had made it himself, and the brightly colored bits of glass seemed symbolic of his devotion to his work.

He oversees a handful of researchers in a lab that is one of about forty on a campus that also includes a vaccine manufacturing facility to support the army's own clinical trials. "They do the work and I write it up. That's all I do is write," he said, referring to both medical journals and the grants that must be written to request funds.

"One of the things that is different about us," emphasized Ann Stewart, "is that our focus is on the adult military traveler. But most of the interest today is on children in Africa, and frankly that's probably what really motivates most of the people who work here."

The military only funds about one-third of the work at Walter Reed. The rest comes from donors or partnering companies.

It would be hard to picture two greater opposites than David Lanar and Stephen Hoffman. Both made their way to the vaunted London School of Hygiene and Tropical Medicine, the Harvard of tropical disease, and joined the military to pursue a passion, but the similarities end there. Lanar is an institutional man, trading independence and potential notoriety for the security and resources the army can provide. Hoffman, neat and trim, politically deft, has to run his own show. Impatient with conventional wisdom and the confines of institutional processes, he is the classic entrepreneur, unleashed, a jay walker, making his own rules as he goes, undeterred by others' definitions of possible and impossible.

"Steve's a smart guy," Lanar told me. "He used to work right here. But what he's trying won't work."

A MASTER SPY'S
MICROSCOPIC TRADECRAFT

"I've always been a bit of a spectrometrist," professor Paul Roepe confided to me in the privacy of his office, just as one might admit, but play down, an unusual fetish or eccentricity. It was his way of explaining how, as a Ph.D. physicist turned molecular chemist and cellular biologist, he'd ended up inventing a revolutionary process for determining how deadly parasites became drug resistant—for actually seeing the treachery their molecules committed inside of red blood cells.

If Steve Hoffman and other vaccine developers are the generals of the malaria battlefield plotting the best strategy for repelling invasion, then Roepe is director of central intelligence. His reconnaissance critically informs where to fight and what weapons to use. Roepe and his colleagues designed the equipment that enables us to spy on the parasite. Though technologically complex, it is based on spectroscopy, which measures the diffusion of light.

Observing the malaria parasite is essential to understanding both how to stop it and how it resists being stopped. The malaria parasite is so tiny that it is extraordinarily difficult to observe. It grows inside of individual red blood cells that have a diameter of about 7 microns. A micron is one ten-thousandth of a centimeter. And the parasite itself isn't even a micron in diameter.

The ability to resist the drugs that are developed to defeat it has enabled the parasite to survive for tens of thousands of years. This ability is shared by other parasites as well as bacteria, tumors, and other diseases. Consequently, the intelligence Dr. Roepe is gathering is coveted by leading medical experts in every field and will almost certainly have long-term applications to cancer, methicillin-resistant staph infections, HIV/AIDS, and the like.

His third-floor office was sandwiched between several small, busy labs at Georgetown University's Basic Science Building. On the door was taped a scrap of white paper that said "2 million children die of malaria every year." Next to it was a copy of an obituary for Arthur Kornberg, a mentor of Roepe's and a Nobel Laureate whose work studying enzymes helped scientists manufacture cells and create the field of biotechnology. On the wall inside were drawings from Roepe's son and daughter, aged nine and twelve.

Dressed casually, Roepe had the lean body of the competitive triathlete that he is. His head was shaved and he sported a two-day growth of beard. He resembled a less menacing version of the actor John Malkovich. He wore a yellow LiveStrong band on his left wrist, and he had used a pen to write a scribbled note to himself on his palm.

When I'd e-mailed Dr. Roepe to request an interview, he had consented but said, "I don't see what this has to do with your work and I'm puzzled about what you think I can tell you of interest." I gathered he was a man who didn't like to waste time.

He certainly didn't waste any during his formal education. His career seems to have followed a meticulously plotted path. He was especially purposeful about pursuing multidisciplinary studies across physics, chemistry, and biology. But serendipity also played a critical role.

I asked if there was any science background in his family:

No, my father was a small town lawyer and judge. But my grandfather was a glass blower. He came here from Scotland. And at the beginning of World War II he realized that the army was going to need syringes and in those days they were all made of glass. So he started making them. That grew into a pretty big glassware products company that supplied a few of the large pharmaceuticals. I remember going to the factory with him and being fascinated by all of the equipment, the glass tubes and beakers and coils. That's when I knew I was going to be a chemist.

After getting his degree in chemistry at Boston University, and then a Ph.D., he did a post-doc at the University of California at Los Angeles and ended up working on tumor drug resistance. He was offered a position at Sloan Kettering in New York, where he worked from 1990 to 1997. He had a corner office with two large windows. By chance it looked out onto the pediatric pavilion where children with leukemia waited for their chemo. "I mean that's what I saw every day. It was right in front of me. All the time. My view was of those kids. That kind of reprioritizes your life. I decided that I wanted my work to be about children, and from there it wasn't far to deciding that it should be children with the diseases that everyone else ignores."

At the heart of the difficulty in combating malaria, as I learned from Roepe, remains a still unknown and perhaps unknowable mystery of nature. This is where Roepe has trained his sights. "Quinine was the traditional drug used to treat malaria, and then came the much less expensive chloroquine, which the Germans created during World War II. But we still don't know exactly how chloroquine works," he told me. "We thought [the parasite] would never be drug resistant, and in fact it wasn't after six months. Instead it took thirty to forty years." Chloroquine's initial advantage over quinine was that it was vastly cheaper, but eventually the parasite evolved to become resistant to both.

The malaria parasite thrives by literally eating the hemoglobin in the red blood cell. What's left as a result of the metabolic process is heme, a toxic substance. To prevent itself from being poisoned by the heme, the parasite is able to crystallize it and sequester it harmlessly off to the side. Malaria drugs interfere with the parasite's ability to do this, but no one knows exactly how. "We try lots of different possibilities until we find a drug to which the disease is sensitive," Roepe explained. "When we find one that works, we go on to solve another problem. We don't spend a lot of time trying to understand why it works."

Just as Roepe was entering the field, a consensus was developing that a certain gene in the parasite was the cause of resistance. Experiments to investigate this multi-drug-resistant (MDR) gene took about ten years. Funding was insufficient, and, according to Roepe, "no one was interested . . . except the NIH, the military, and the Brits." It was Roepe's experiments that ultimately disproved the theories about an MDR gene. "The idea that one gene could be responsible for resistance is a gross oversimplification," Roepe said. "There were a lot of people unhappy with me for coming in as a young upstart and making this claim. Today everyone agrees with it but some are still unhappy with me. I guess I've always been a renegade."

Though Roepe has dramatically advanced the field's knowledge, and is on the cusp of being able to help unravel the age-old mystery of how drug resistance develops, he didn't foresee being able to completely solve the drug-resistance conundrum. "I think of it as staying ahead of the resistance curve," he told me. "The parasite will continue to mutate and adapt and we will always have to develop new drugs in response. But it used to be very difficult to know which drug to use because it was difficult to know which drug the parasite would resist. Now a blood test can tell us this almost instantly. That gets the cost down." Which is a critical factor in underdeveloped regions of the world. As Roepe recalled a military corporal in Southeast Asia once telling him, "you've gotta make it for 50 cents a dose or you might as well not make it."

When the existing technology was too limited for Roepe, he invented new technology. When the knowledge in one field of science was insufficient, he collaborated across disciplines, bringing in physicists and molecular biologists. When he concluded that existing diagnostic tools enabling doctors to match medicines to the type of malaria were too time consuming and expensive, he made economics the driving force behind creating a better method.

"With all due respect to Steve Hoffman, a vaccine would be great, but that's at least ten years away," said Roepe. "And with 2 million kids dying every year from malaria, that's 20 million freakin' kids that will die," he added, his voice starting to rise. "In my humble opinion the Gates Foundation ought to balance a bit more of its funding to get drugs to these kids now."

The kind of work Roepe is doing gives malaria drug developers their best chance of keeping pace with the parasite's relentless ability to evade their attacks. And he put his finger on the classic tension that continues to exist between those who would invest in long-term efforts to actually eradicate the disease, like a vaccine, which may seem impractical and far-fetched, and those who believe that the pressing nature of urgent human needs demands more immediate action. Faced with the reality of finite resources, it does not always seem feasible to do both. But that's just the kind of failure of imagination that Steve Hoffman has fought to overcome.

A CLUB MED FOR BUGS

The Bloomberg School of Public Health on the Johns Hopkins University campus in Baltimore is home to one of the world's most sophisticated insectaries. It is not some sort of showcase for insects, however, with exotic, winged creatures buzzing and flying about. Instead, it exists for a sole purpose: breeding mosquitoes for laboratory research. And when you first pull open the heavy door and walk in, it is so quiet and still that you get the misleading impression that nothing is going on.

The lab at the school's Malaria Research Institute actually includes seven separate insectaries producing thousands of mosquitoes a week. Each insectary is accessed through a large steel door that looks like the door of a walk-in freezer—except that the insectaries are kept warm inside and at 80 percent humidity, to recreate the climate in which mosquitoes breed. A computer controls the lighting, which mimics sunrise and sunset. Insectary mosquitoes are fed their favorite foods and encouraged to mate, and the weather is perfect. It is Club Med for bugs.

Johns Hopkins built the lab when it launched the Malaria Research Institute in 2001. (The school also established a site in Zambia's Southern Province for field research.) The staff, to a person, harbors an ardor for their work that may seem unusual to outsiders. "A flagellating protozoon is very, very beautiful," said one professor, who also works on sleeping sickness caused by Trypanosome parasites.

Another, David Sullivan, is an assistant professor of molecular microbiology and immunology and has been working to develop a simpler diagnostic for malaria that would not require drawing blood. When I asked him why he became so interested in malaria, he responded, "What I don't understand is why everyone else isn't interested in malaria. I realized it was the one thing I never minded waking up at two o'clock in the morning for, to run down to the lab, to check on an experiment."

The Bloomberg School of Public Health is the largest school of public health in the world. Philanthropic generosity enabled it to create a lab second to none. "This is luxury as far as labs go," explained Marcelo Jacobs-Lorena, a professor of molecular biology who was lured away from a twenty-six-year career at Case Western Reserve. He smiled broadly as we surveyed a long rectangular room of work benches, microscopes, computers, and shelves crammed full with beakers, bottles, and test tubes. Several of his students were at work, and as modern as this lab may be, they were still doing science as good science has always been done, patiently extracting liquid from a vial, mixing it, observing, and recording results.

Like many others at the same senior stage of their careers, Jacobs-Lorena has found that his administrative and grant-writing responsibilities keep him from conducting experiments himself. But he lit up when we walked into the lab and clearly enjoyed telling me about the work that was underway there. He began by showing me how the researchers put anesthetized mice on top of the mosquitoes' wire-mesh cages so that the female mosquitoes could take the blood meals they required for the protein needed to lay eggs.

The eggs are collected on a thin, soft pad and then placed in an uncovered white tray filled with water, where they lay until they hatch as larvae. Tray is stacked upon tray in a large metal rack, like cookie sheets at the bakery. Dark specks float near the surface or cling to one side. The larvae are fed on pellets of cat food that become bloated and float in the liquid, and then the larvae become pupae. Near the end of their ten-day journey to adulthood, a cloth is placed over the tray so they won't be able to fly away. The mosquitoes are then vacuumed into another cage where they await the microscope or the dissecting blade of a student or post-doc. Their life span of thirty days lies ahead.

There are elaborate precautions to keep the mosquitoes confined to the insectary. In each room there was a bug zapper, a lighted blue tube reminiscent of the contraptions I'd encountered in Hoffman's offices. At this lab, however, there was a net affixed underneath each tube. While Jacobs-Lorena was talking, I noticed with surprise a mosquito zipping around my head, and instinctively, I swatted at it. "There's always one that seems to get out," Jacobs-Lorena said, more with amusement than alarm. Every wall contained posters with emergency procedures and phone numbers to call should there be any kind of accident.

As we entered the special insectary, which is treated as if it were biohazard level 3, he explained that this was where they dealt with the malaria-causing *Plasmodium falciparum* parasite. Because only female mosquitoes bite and spread malaria, the males are useless to the lab and must be separated out. I asked Jacobs-Lorena how they could tell them apart. He said to just take a look. To my eyes they were blurry tiny black bugs, no larger than a sprinkle on a cupcake, and indistinguishable from each other. But once he pointed out that the males were long and slender and the females were thicker, with a bulge in the middle, I was able to notice the difference even in the small specks clinging to the containers. "Cold knocks them out and makes them lie very still, so we basically make them cold and then put them on the lab bench and separate the males and females by hand," he told me. It's a strange labor of love.

After that, they are injected by a fine needle with foreign genes in an effort to make them incapable of transmitting parasites. The goal of Jacobs-Lorena's lab, and his work since 1982, has been the development of a transmission-blocking vaccine. His experiments are designed to determine whether mosquitoes can be genetically modified to interfere with the parasite's invasion of its midgut and salivary gland. It is in the midgut that the sexual form of the parasite develops, and in the salivary gland that it waits on deck to infect the next human that is bitten. Unlike Hoffman's vaccine, Jacobs-Lorena's would not attack the parasite in the liver or anywhere else in the human body; it would instead make it impossible for the parasite to travel by mosquito, and therefore unable to infect humans. Think of it as a genetic modification that takes away the parasite's car keys.

To get into the mosquito's midgut or salivary gland, the parasite must cross through a membrane called the *epithelium*. Not much is known about how it actually does so, but it's more complicated than boring right through and in. Jacobs-Lorena described it as more like a key turning a lock; getting the combination just right depends on the molecules found on the surface. Making adjustments to those molecules to prevent entry is what Jacobs-Lorena's work has been about. It is criticized by some who say he is working to protect mosquitoes from infection, rather than people, but of course, if mosquitoes weren't infected, people wouldn't be either.

In 2000, Jacobs-Lorena pointed out in *Parisitology To-day* that "among the five most deadly infectious diseases (acute lower respiratory infection, TB, cholera, AIDS and malaria) only malaria requires a vector"—or channel (in other words, the mosquito)—"for transmission. In the past, campaigns to control mosquito populations have resulted in dramatic decreases of malaria incidence. However, insecticide resistance and environmental damage quickly reversed early successes and now complicate vector control."[6] As a result, techniques such as manipulation of vectorial capacity—meaning manipulation of the molecules inside a mosquito—are now being considered.

Jacobs-Lorena acknowledges that his is just one of many competing approaches to controlling malaria, and he knows it has its detractors. But in many ways each of his competitors is in the same boat, facing the same dilemma. Searching for more and more obscure clues, and testing theories that seem more and more eccentric, each has his own hypothesis, one in which he has invested his time, talent, and reputation. And each keeps a wary eye on the progress of his peers.

Just down the hall from Jacobs-Lorena is the laboratory of Nirbhay Kumar, who came down with malaria as a young man studying for his Ph.D. in New Delhi in 1976. He had always planned "to go back to India and do good science there," but then a mentor persuaded him of the contribution that could be made on malaria if he stayed in the United States. He is now coming up on his twentieth year at Johns Hopkins.

Kumar has been working on a vaccine to prevent transmission by mosquitoes, but in his case, it would be by blocking the sexual development of the parasite within the mosquito. In this way Hopkins has carved out a research niche for itself.

Kumar explained why he has chosen to work on disabling the parasite: "There are only three variables for researchers to attempt to control: the mosquitoes, the parasites, or the people. You don't want to eliminate the mosquito. They play other roles in nature that we need. And you don't want to eliminate people. That leaves the parasite."

If Steve Hoffman's approach is akin to an army general seeking to contain the enemy on the battlefield of the liver where it will be trapped and perish, Kumar and Jacobs-Lorena are focusing on disabling the enemy's air force, or at least the air force's munitions capability. Mosquitoes are the aircraft, delivering their deadly payloads with awesome precision. It's not the vehicle that the does the damage, but what it carries. And if the vehicles can be modified so the munitions are disarmed before they are released, then we will have nothing to fear.

There are many other approaches being developed as well. In addition to the thirty-five different malaria vaccine development efforts being monitored by the World Health Organization, there are also researchers seeking to find curative drugs, efforts to distribute insecticide-treated bed nets, and other means of fighting malaria. There are two reasons for this vast variety of efforts and approaches.

First is the very complexity of the parasite's development and life cycle and the many moments of potential vulnerability that scientists see as targets to strike. As it bodysurfs blood from one human victim through the mosquito to the next human victim, the parasite morphs and differentiates into six completely different forms. It goes through fertilization and then invades various organs and cells.

Different efforts to combat the disease are aimed at different moments in this cycle, but if we are to prevail, we may eventually have to fight the parasite at every point. The paristologists have essentially adapted Winston Churchill's famous World War II exhortation: "We shall fight on the beaches, we shall fight on the landing grounds, we shall fight in the fields and in the streets, we shall fight in the hills."

As Kumar put it, "we have a problem of multiplicity." Researchers must confront multiple species of the parasite and multiple strains of each species. The parasite itself has multiple life-cycle stages, invades multiple strains of the mosquito, and exists across multiple epidemiological areas. And most challenging of all is that there are multiple immune responses. Kumar's colleague David Sullivan captured the daunting challenge when he told me that any vaccine developed "has to be better than nature."

Second, because the Plasmodium parasite has evaded or defeated every effort to destroy it, scientists have been forced to pursue increasingly radical, seemingly unrealistic options. All of the most logical ideas have already been tried, and they have failed. What remains are the less logical ones, certainly the less obvious ones.

All the researchers I met acknowledged being sobered by the experiences of their predecessors. They live with the reality that their efforts are only one piece of the puzzle, with success ultimately being dependent upon factors beyond their control—such as, for example, government and public policy.

Teresa Shapiro, professor of clinical pharmacology at Hopkins, explained her job to me as "developing the tools. That's what I have to keep my focus on." She noted that "success in fighting malaria has waxed and waned over the years, often set back somewhere because of the loss of political infrastructure." "If politicians and government officials decide they want to deal with malaria," she said, "my job is to make sure the tools are available. It's about a lot more than whether we come up with an affordable vaccine. As Jean Genet said at the World Health Organization, 'even free is too expensive,' because the costs are in distributing it."

"VACCINE MAKERS, LIKE OIL MEN AND GOLD DIGGERS, NEVER STOP DREAMING"

In the fall of 2006 I drove with some trepidation out to Rockville to reconnect with Steve Hoffman. The last time I had seen him, he'd been so stressed about his biotech business running low on funds that he had worried about having to shut down Sanaria's operations. I had been checking the company's website often, and there hadn't been an addition to it in months. No new press releases, either. Nothing surfaced through a Google search that I hadn't seen before. Most unusual, there were no new papers written by Hoffman, who has always been one of the most prolific authors in the field of malaria research and tropical medicine. He had spoken of a desire to have a new facility, but his ability to afford one would depend on receiving a major grant. I was worried.

I didn't think of Hoffman as someone who gave up easily. Quite the opposite. He is fiercely competitive and proud. Maybe sounding the alarm about finances was just his way of calling the bluff of a prospective donor. Perhaps he'd built in a financial cushion and there was really nothing to worry about.

Although Hoffman had experience raising funds and managing a department in the navy, he had never run a business, and he had no financial training. For the most part, his career had been lived safely in the confines of enormous, well-funded institutions like the U.S. Navy and Craig Ventner's Celera. It was usually someone else's job to worry about the money, meet payroll, manage staff. All Hoffman ever had to do was think about the science, design experiments, and publish research results. Considering that everyone else I'd talked to was skeptical that Hoffman's vaccine, based on building an assembly line for dissecting mosquito salivary glands, was highly impractical, I started to wonder if he and the lab he'd cobbled together might disappear without a trace.

Science is all about trial and error, advances and setbacks, and meticulous preparation dedicated to discovering the needle in the haystack. It makes the lab a fascinating and exciting place, but the rollercoaster ride takes a toll on the human beings who perform the experiments. Emotions inevitably seep into even the most sterile labs. Disappointment, anxiety, and fear can never be distilled out completely. I'd detected trace elements of them in Hoffman—mostly in some worried comments he'd made about Sanaria's precarious finances—and wondered how he'd respond.

When I got to his office, he was on the phone with a scientist in Colombia. I sat a few minutes looking at some medical journals and admiring his artwork. His walls were covered with more African sculptures and photos than I remembered from previous visits. The place had much more of a "lived in" feel to it. Since I was sifting for evidence that he and Sanaria were here to stay, I seized on this as a good sign. It turns out that I didn't have to search much farther.

After concluding his call, Steve stood up and came over to the small round conference table where I was sitting. For the first time since I'd met him, his smile seemed more natural than forced. I asked for an update on where he was now with funding, toxicity trials, production schedules, and the new facility he planned to build.

"July and August were the worst two months of my professional life," he began, but with the grin of a man reporting on his own eventual triumph. "Waking up at four in the morning, with that pit in the stomach over whether we were going to go negative, whether we'd have to stop operations. I'll never let myself get in that position again."

Although it was only a fraction of what he needed, the Malaria Vaccine Initiative (MVI) had come through in July with bridge funding: $534,000, and then an additional $200,000—enough to keep the operation going. MVI was created by a Seattle-based organization called PATH that was started in 1977 and sought to advance technology to improve health. MVI was funded by Bill Gates to identify the most promising vaccine candidates, accelerate their development, and ensure eventual licensing and accessibility in the developing countries where they were most needed. They currently fund ten separate vaccine-development efforts around the world.

And then Hoffman received the best news he'd heard in a long time. "On August 11, while my family and I were vacationing in Alaska's Denali National Park, I received word that $29.3 million in Gates Foundation money would be granted through MVI," he said. "We toasted each other over the phone, celebrated, the whole thing, and then a week later they came back and said there were problems with the details of how the grant had been written and we had to go back and start working on it again."

The funding eventually came through: "It took a lot of work," Hoffman said, "but the Gates funding changed everything. It's not just a matter of being grateful. Many of us would not be doing what we are doing today if not for the Gates Foundation." The funding would enable him to build a vaccine manufacturing facility and keep going for at least three more years.

"But now that we've got the funding I've actually had a kind of postpartum depression, just some feeling of letdown after all of that time," Hoffman told me, recounting recent experiences:

> We presented our work at the American Society of Tropical Medicine and Hygiene conference in Atlanta. Our family had been in Singapore with Ben and Seth, who were competing in the World Kung Fu Championship, and then Alaska. So I was jet lagged. I also picked up some viral infection. I've been at every ASTMH conference since 1978 and there were 2,500 participants at this year's, and I know everyone, but I ended up having dinner alone and then going to my room to rent a movie. I had some weird kind of depression now that the money had come through.

Still, the conference was a success in some significant ways, and Sanaria was still making progress: "I think it is fair to say that people at the ASTMH conference were blown away by the results we showed," Hoffman said. Sanaria's was "the only vaccine with a high rate of effectiveness over time. None of the others show that. None of the others come close. Melinda Moree [the director of MVI] introduced me and it was a very nice introduction."

Hoffman emphasized that he only gave part of the presentation and the rest was by his team, whose growing size and diversity further impressed the attendees as an indication of Sanaria's financial health.

I asked Hoffman who his skeptics were at the Atlanta conference and he described them as "the GSK/Walter Reed axis." GlaxoSmithKline and the Walter Reed Army Institute of Research have had a long partnership developing a vaccine. Hoffman is publicly polite about their effort, but privately skeptical, especially of how they report their clinical trial results.

He explained that his critics were skeptical that Sanaria would be able to freeze the vaccine in liquid nitrogen, a step that would be essential to storing and transporting it. Just talking about it brought out Hoffman's competitive streak. "First they said we couldn't get enough sporozoites, then they said we couldn't do it and keep it sterile," he said. "Then they said we couldn't keep it stable. And we proved them wrong on all of those. So now they're saying there will be problems with the liquid nitrogen."

We talked about what comes next now that the money had been secured. The immediate focus was on building the new manufacturing facility in Rockville. It would have to meet the FDA's GMP standards for Good Manufacturing Practices. The other major goal was to finish the toxicology studies and then begin clinical trials.

In one sense, Hoffman had cleared the highest hurdle of funding—getting that first big break and the recognition that came with it. But in another he had graduated to a sweepstakes where the odds were even more daunting. The Food and Drug Administration's drug-approval process is an obstacle course and endurance trial so Byzantine, costly, and time-consuming that, while many enter, few come out the other side. Gaining drug approval makes running for president look like an inexpensive walk in the park.

"We should submit our IND by the first quarter of 2008," Hoffman said, referring to the mandated Investigational New Drug Application Process. The IND shows results of previous experiments; how, where, and by whom the new studies will be conducted; the chemical structure of the compound; how it is thought to work in the body; any toxic effects found in the animal studies; and how the compound is manufactured. It becomes effective if the FDA does not disapprove within thirty days.

After phases one, two, and three of clinical trials, a drug being developed finally files a Biological License Application, which provides data attesting to the safety and effectiveness of the vaccine. It typically runs more than 100,000 pages. Statistics about drug approval in the United States give an idea of the long odds Hoffman faces:

- Only 5 out of every 5,000 compounds that manage to enter preclinical trials make it to human testing.
- The average length of time it takes to get a drug from lab to pharmacy shelf is twelve years, and the average cost to get it there is more than $1 billion.[7]

Of the many labels I'd used to describe Hoffman— physician, naval officer, research scientist, fundraiser, evangelist, biotech engineer—I now added one more: gambler.

Steve Hoffman not only has to bet on the long odds of getting his vaccine from lab bench to pharmacy, but then must place a second bet that he'll get there ahead of all the competitors working on different vaccines that may prove more effective or affordable.

The nearly $30 million from Gates and via the Malaria Vaccine Initiative allows Hoffman to cover his bet, but nothing more.

His work, like that of so many of his rivals, still depends on what idealism, imagination, and persistence can achieve in that narrow but vitally important space between the impractical and the impossible. It is a space I have come to think of as the imagination gap.

The imagination gap is a place where hope lies waiting to be discovered, and cannot be extinguished once it has. Most failures in life are not failures of resources, or organization, or strategy or discipline. They are failures of imagination. It can be a very lonely space, where one faces skepticism and even ridicule, often from one's own colleagues, a place that one can escape from intact only with sufficient reserves of confidence, stubbornness, and steely resolve.

CHAPTER 7

THE E WORD,
ONCE AGAIN AND AT LAST

British scientists have pioneered a vaccine against malaria that they be-
lieve could save millions of lives. . . .

Adrian Hill, a professor from Oxford University's department of medi-
cine, who is heading the project, last night confirmed the breakthrough. . . .

"What is different about our approach is we are trying to stimulate a
different arm of the immune system as the parasite lives inside the cells
most of the time," Prof Hill said. . . .

The new approach promotes the production of vital immune cells
known as "T cells" in the body, which then destroy malaria infected cells.

—John Schutzer-Weissmann and Lorraine Fraser,
"British Vaccine Breakthrough Will Save Millions
from Malaria," *The Telegraph*, August 18, 2002

A VACCINE TO PREVENT MALARIA is a wonderful idea that would save millions of lives, but it is the Holy Grail of tropical medicine. So far, no one has been able to find or attain it. Military labs, giant pharmaceuticals, and thousands of brilliant researchers have spent hundreds of millions of dollars and as many hours trying everything possible but have not yet succeeded in producing a vaccine that is safe, effective, and reliable. And even if a vaccine is discovered, there may always be children who slip between the cracks and don't get it, allowing the parasite to live to fight another day.

Even though—to date—it can't be prevented by vaccine, malaria can be cured if treated in time. Doctors can only save people if they can diagnose the disease accurately and if they have access to needed medicines. Both are easier said than done in the developing world, where labs, equipment, and pharmacies are scarce. The drugs are still often too expensive for those who need them the most. And so, in a parallel universe to the vaccine development community, researchers also work on developing a drug that can be produced on a large scale at low cost.

Three thousand miles from Steve Hoffman's lab at Sanaria, on the opposite coast, Jay Keasling, a professor of chemical engineering at the University of California at Berkeley, is working to advance a medicinal cure instead of a vaccine. His strategy is based on a form of synthetic biology that could dramatically reduce the cost of malaria treatment. If either Hoffman or Keasling achieves a triumphant breakthrough, funds for the other could dry up. Which means that Hoffman is not only competing against other vaccine developers, but against Keasling and a large number of others in the medical establishment who believe that the more direct route to saving lives is through the proven track record of drugs. For many of them, a malaria vaccine represents nothing more than unfulfilled dreams that have little chance of becoming reality.

A coworker of Keasling's at Berkeley's Lawrence Berkeley National Laboratory, Lynn Yarris, has put it bluntly: "The complex life-cycle of *Plasmodium falciparum*, the parasite that carries malaria, makes it impossible to eradicate the disease. Treatment is the only option and the most effective current treatment is artemisinin."[1]

Artemisinin is the one drug that remains effective against malaria strains that are now resistant to front-line drugs like chloroquine. The Chinese have been using it as an herbal remedy for more than 2,000 years, but it was the Vietnam War that provoked them to explore its properties regarding malaria. It has a nearly 100 percent success rate for all known strains of malaria, destroying the parasite by releasing high doses of oxygen-based free radicals that attack the parasite inside the iron-rich red blood cells. The controversies that surround the varying degrees of effectiveness of malaria vaccines in trial are not relevant to artemisinin. Artemisinin works (although indications of drug resistance have suggested the need to deliver it in the form of combination therapies with other drugs rather than by itself).

It only came to the attention of Western researchers by chance in 1979, when Nick White, the director of the Southeast Asia office of The Wellcome Trust, one of the largest medical research foundations in the world, received a dog-eared copy of a paper from a Chinese medical journal from a friend in Hong Kong. It reported the results of clinical trials conducted by a team of scientists created by Chairman Mao for the express purpose of screening herbal remedies for a malaria cure. "Gobsmacked," White arranged to visit the scientist who led the team, Professor Li Guo Qiao, and when he left that meeting, he had a bottle of artemisinin in his hands.[2]

But artemisinin is not easy to get hold of. It is extracted from the dry leaves of the sweet wormwood tree, a six-foot-tall plant that can grow in many places but only produces artemisinin in the mangrove swamps of China and Vietnam. Not only is the plant rare, but the substance is difficult to extract, which makes it expensive. The process is labor intensive, sometimes involving diesel-fuel purification methods that can leave toxic impurities in the final drug product. There have been reports of speculators buying and hoarding the plants, quadrupling the price and making it even less accessible to those who need it. In the fall of 2005, prices went from $115 a pound to more than $400. By 2008, additional planting on the part of farmers had reduced the price to $70 a pound.[3]

Still, the World Health Organization predicts that for the foreseeable future, demand for artemisinin will continue to greatly outpace supply. Keasling expects a shortfall of at least 100 million treatments in 2010 and 2011. Kent Campbell, director of the Malaria Control Program at PATH, underscored the stakes in an interview with *The New Yorker*: "Losing artemisinin would set us back years, if not decades. . . . One can envision any number of theoretical public-health disasters in the world. But this is not theoretical. This is real. Without artemisinin, millions of people could die."[4]

A course of treatment for adults, produced by Novartis, comes to about $2.40 per person. Novartis has already been accused of breaking commitments to supply sufficient amounts to meet projected demand. Médecins Sans Frontières, the international humanitarian organization also known as Doctors Without Borders, argues that the company is not interested in a market that is not profitable. In 2005, Novartis president Daniel Vasella, speaking to the *Financial Times*, said, "We have no model which would [meet] the need for new drugs in a sustainable way. You can't expect a for-profit organisation to do this on a large scale."[5] He was referring to the lack of paying customers through which pharmaceutical firms could make back the investment necessary to produce sufficient quantities of the drug.

Left to its own devices there is simply no way nature will produce enough of this critically needed and already proven cure. But Jay Keasling is quite as capable as Steve Hoffman of making a leap of imagination: "I see no reason why we can't completely reimagine the chemical industry," he told me.[6] "With the tools of synthetic biology, we don't have to just accept what nature has given us." It's a philosophy so bold as to border on hubris, but with a million kids a year dying from malaria, Keasling's drive to find shortcuts is fueled by a sense of urgency. Keasling is using genomics to accelerate nature's own processes and manipulate them to create more of the products we need.

Keasling was raised on his family farm outside of Lincoln, Nebraska, and synthetic biology was unheard of when he took his first genetics class at the University of Nebraska in 1983. But by the time he reached the Ph.D. program at the University of Michigan, he "wanted to manipulate a cell like an engineer does a chip," he told *California* magazine.[7] Following a post-doc at Stanford, he arrived at UC Berkeley's Department of Chemical Engineering in 1992 with some unique ideas about reengineering enzyme reactions within microbes.

Just over ten years later, working from home just as Hoffman had, Keasling launched the biotech company Amyris to carry out those ideas. He has not left the institutional realm entirely, like Hoffman has. Instead, Keasling has deftly merged several roles, not only heading up Amyris Biotechnologies but also continuing to teach chemical engineering and bioengineering at UC Berkeley and serving as founding director of the Synthetic Biology Department at Berkeley, CEO of the Joint BioEnergy Institute, and acting deputy laboratory director of Lawrence Berkeley National Laboratory.

Keasling aims to make it much cheaper to produce artemisinin, thereby bringing down the price and making it possible to treat the vast number of malaria victims who now go untreated. Speaking with a reporter from the UC Berkeley News in 2004, he explained how, saying, "We are designing the cell to be a chemical factory." Like Hoffman, Keasling turned to business because he saw the potential for commercial enterprises to bring an idea to scale. Capitalized by a grant from the Gates Foundation, Amyris uses synthetic biology to isolate genes from their natural sources and insert them into industrial microbes, and thereby to produce natural compounds much more cheaply.[8]

Keasling's research team mixed different genes from different organisms to perform chemistry inside living cells. They combined genes from yeast and the wormwood plant with the common intestinal *Escherichia coli* bacteria cell to create a new metabolic pathway—a series of chemical reactions that occur within a cell—that yields a form of artemisinin called amorphadiene. Amorphadiene is an artemisinin precursor that is already used as the basis for all of the artemisinin drugs currently on the market. Keasling and his team believe they are on track to produce 10,000 times the amount possible by the old ways.

Thus Keasling's team encouraged the bacteria to produce molecules that are not found in sufficient quantities in nature. Metabolic engineering of microbes co-opts the microbes' metabolism for our own benefit. It could replace the expensive techniques used in the chemical industry today—not just for making artemisinin, but for making a wide variety of other drugs that are badly needed but difficult to produce in the quantities needed around the world.

Keasling believed the artemisinin he created in his lab could be sold for one-tenth of the price of what other producers were charging, meaning about 21 cents a dose. "We're taking a natural product in short supply, using biotechnology to produce it and to produce it very inexpensively in the developing world," he told me. Providing it to 70 percent of the malaria victims in Africa would cost about $1 billion.

As different as Hoffman and Keasling may be in terms of their approaches to malaria, they both work at the intersection where science, philanthropy, and entrepreneurship are converging. They both are taking a known, effective solution and trying to make it affordable and sustainable. Their innovation is to produce it at a scale that drives down cost, and therefore price, and to do so by allowing nature (the mosquito, or the enzymes) to do the work rather than manufacturing it at great expense.

Jay Keasling and Steve Hoffman have met only once, at a conference where they both happened to be giving a talk. Though they express genuine admiration for each other's accomplishments, each is so focused that, even when prompted, neither shows much interest in the other. But when I first started visiting Keasling I could see that he was philosophically connected to Hoffman in ways beyond their shared commitment to ensuring that children didn't die from malaria.

The first thing Steve Hoffman told me on the first day we met was that the solution to a malaria vaccine was not a matter of scientific discovery but rather of biotech engineering. Though he was overstating the point, because science played a critical role at every stage of his effort, he was trying to convey that we knew what worked, but the challenge was whether we could scale it. In leading the effort to apply engineering principles to biology, Keasling demonstrated a nearly identical insight in terms of what it would take to make highly effective medicines available on a mass scale, the kind of scale that would ensure that children like Alima were quickly cured.

Both were producing their product in a place that literally seemed impossible right up until they did it: Hoffman using the insides of mosquitoes as his primary production facilities, and Keasling using cells as chemical factories. They not only managed to see what the naked eye could not, but also to imagine a possibility that no one else had imagined. Both were undeterred by either the complexity and unknowns inherent in such virgin territory or the skepticism of respected colleagues. The conviction that they could save lives offset the risks they would be taking.

At the end of 2004, the Gates Foundation awarded a $42.6 million grant to support Keasling's work through a partnership between UC Berkeley, Amyris, and the Institute for OneWorld Health, the nonprofit pharmaceutical created in 2000 to help develop drugs and vaccines for neglected diseases. Keasling's responsibility was to do the research necessary for perfecting the microbial factory for artemisinin. Amyris would develop the process for industrial fermentation and commercialization. OneWorld Health would handle the regulatory work.

It was expected to take five to ten years to develop the microbes into truly large-scale producers of the artemisinin precursor. Keasling also hoped to produce promising anti-AIDS drugs and cancer-fighting medicines like Taxol using the same processes.

A CHANCE TO
SAVE THE WORLD—TWICE!

The offices of Amyris overflow like a lab flask that can't contain the chemical reaction within.

The firm spills across a courtyard into a second building at the edge of Emeryville, California, near Oakland. There are nearly a hundred staff members, including biologists, chemists, and newly hired administrators and project managers.

Producing an inexpensive antimalaria drug is not the only project that consumes their attention. Amyris takes aim at another, very different global challenge as well, and it exploits the same pioneering technology. Close on the heels of the effort to create synthetic artemisinin, and in danger of overtaking it, is the effort to produce synthetic fuels. If the project is successful, it could reduce our dependence on oil and mitigate global warming.

Keasling's insight, simple in concept but extraordinarily complex in the lab, is that rare and valuable plant-based substances could be grown in lab tanks. All you have to do is transplant genes from the plant into fast-growing bacteria like *E. coli* and yeast. The insight has a lot of different applications, most of which are related to medical cures. But, theoretically, it could also be used to produce a next generation of carbon-neutral biofuels.

Amyris's founding partners, while motivated by a desire to help malaria victims, were also savvy enough to appreciate the potential for enormous profit that lay in the field of synthetic biology and the processes they were perfecting. But what animated their conversation and sustained absurdly long hours in the lab was the thrill of basic science: the prospect of discovering something never before known, of peeking behind nature's curtain to get a glimpse of the universe's secrets, of making real to others what once they alone imagined.

Jack Newman is the forty-something vice president of Amyris overseeing lab research. The son of peripatetic artists, he dropped out of high school at fourteen. By his fifteenth birthday, he was attending community college, and en route to a degree at UC Berkeley in molecular and cellular biology. Pursuing a doctorate at the University of Wisconsin, Newman heard Keasling guest-lecture and determined to do his postdoctoral work in Keasling's lab

"My life is a movie," Newman told me. "Even I can't believe it turned out the way it did. Amyris is exactly the company that I wanted to work for from the time I was fourteen. It is everything I imagined and I've never wanted to do anything else. It is exactly that company in every way."

In black jeans and a black T-shirt, with long black hair flowing behind a prominent forehead and beyond his shoulders, he looks more rock band or computer geek than Dr. Newman. Some of the company's energy seems bottled up inside his body, which moves in awkward ways, his head occasionally tilting then snapping back. A natural teacher whose enthusiasm is infectious, he jumped up to the white board to diagram for me their process of using chemistry to create amorphadiene and artemisininc acid.

"The chemistry isn't simple, but it is reliable," Newman said. "You get an outcome you can count on. Biologic processes don't work that way. Historically biology was too complex to apply engineering to, but that is changing." Ever faster genome sequencing has made it so.

At the time I visited Amyris, the milestones that Amyris, Keasling's lab, and the Institute for OneWorld Health had agreed to with the Gates Foundation focused on getting the metabolically engineered microbe to produce artemisinin on the scale of 25 grams per liter. At first they were getting 0.0000000001 gram per liter. Twenty-five grams per liter would be far more than has ever been produced in a lab, and a telling indicator of future potential. They've been testing both yeast and E. coli. "Both crossed the finish line," said Newman of the ability of yeast and E. coli to grow large-scale batches of the drug. "Just this morning an experiment on this came back with results that were beautiful."

Newman didn't use "beautiful," to mean "great" or "cool." He meant beautiful the way it is used to describe a Monet or a Matisse: elegant, balanced, pleasing to the senses. He looks at chemical analysis and sees art. The 25-gram mark had not yet been reached, but things looked very promising.

Keasling walked me through room after room of tubes, tanks, coils, and monitors, most of it connected to enormous computing power, young technicians, some with arms covered in tattoos from wrist to shoulder, strove to create artemisinin in ever greater quantities.

Just as Steve Hoffman would wrestle with whether what worked via the bite of the mosquito would work when scaled into a clinical manufacturing process, Amyris faced the issue of whether what worked in the shake flask on a lab shelf would work in a 50,000-liter tank the size of a bus. Scale changes chemistry. For example, the pressure at the bottom of the tank is different from the pressure at the top. That alone can completely alter results. And there are different surface-to-mass ratios, which vary with exposure to oxygen.

Amyris's president, Kinkead Reiling, fell in love with math and science at a young age. In blue jeans and a button-down striped shirt with a neatly trimmed goatee, he seemed like an earnest, cautious person. His eyes lit up as he described a protein that, if pulled taut, would be like a long string of spaghetti, and when let go would fold up and in on itself.

Reiling, a military brat whose father flew cargo planes for the U.S. Air Force, started out in physics at the University of California at Los Angeles, fascinated by the notion of being able to understand the world by understanding the most basic parts of which it is made. He switched to biology at Columbia University and had the chance to work with Professor Robert Stroud, whose legacy was great advances in protein crystallography. He's been a jack-of-all-trades at Amyris, reluctantly moving farther from the science to assume business responsibilities for which he has no training. Finances, office space, human resources, hiring, payroll—all of this is Reiling's portfolio. Akin to the oldest child in a dysfunctional family, he had to be the responsible one.

An added undercurrent of complexity runs through this company because its investors have varying expectations and objectives, even though their interests may be aligned. When Bill Gates funded Amyris to develop artemisinin, there existed essentially no commercial market or likelihood of financial gain. But in 2006, venture capitalists Kleiner Perkins and Khosla Ventures put $20 million into Amyris and arranged for John Melo, former head of fuel operations for BP, to become president. Shortly thereafter BP committed to invest $500 million over ten years into a new Biosciences Energy Institute at UC Berkeley. BP, looking ahead to whatever might end up being the next generation's source of energy, apparently sees the potential for profit in the development of an alternative to fossil-fuel extraction.

Keasling has little time for what goes on outside of Amyris. At a 2007 Harvard School of Public Health conference at which he presented, he encountered—but did not really meet—"the vaccine guys." He not only doesn't know Steve Hoffman from Sanaria and Rip Ballou from GlaxoSmithKline, but doesn't seem to focus much on what they are up to. They are the other side of the ideological wall: the hopeless idealists pursuing a total victory over malaria while losing countless small wars in the meantime. Keasling is busy fighting the small wars: "They've been at it for fifty years and there has still never been a vaccine to reach the market, while meanwhile millions of kids die each year. So we better get some medicines to those kids."

His schedule could compete with that of any other driven person I've ever met: "I wake up every morning at 4:30, it's the most magical time of the day, and I go to the gym from 5 to 7. Then I'm at Berkeley, teaching." He has about fifty students. "And in the afternoon I get over to the lab, and a couple days a week to Amyris. There is some complaining about not being able to get on my schedule, but I don't want things on my schedule."

He said teaching is the most important thing he does, but also admitted, with a somewhat pained expression, that he may not be teaching forever. It depends on the grants he gets.

Describing the decision to take VC money and then BP's investment, and what it meant to give up control, he was clear: "If the question is, Do you want to own 100 percent of zero, or 20 percent of something really big, I know what I want."

Everyone I spoke with at Amyris sounded a bit defensive about the dual mission of combating malaria and inventing biofuels, but they all also went out of their way to praise John Melo, who came in as Amyris's CEO after a career at BP and Amoco. Reiling described him as "the least oily oil guy" he knew. "He really gets what we're doing. Doesn't totally understand the science, but respects it." And he also made the point that all interests are aligned. "Amyris only works if our work with artemisinin is a success," Reiling said.

Both Keasling and Amyris were in discussions with the U.S. Department of Energy about grants that would further their work. And like Hoffman at Sanaria, Keasling's team was giving some thought, even though it is technically beyond their sphere of influence, to how the medicine will actually be distributed in Africa. "Everyone here is motivated by seeing the science actually used," Keasling told me.

The example of Amyris makes a compelling case for nonprofits embracing the disciplines of science rather than fixating only on the need for MBAs to secure the "entrepreneurial" label. The team at Amyris is trying to solve almost unimaginably complex social problems through biotech engineering. They are trying to scale up proven solutions and to lower their cost.

Scientists bring proof and precision to their efforts. Clinical trials guarantee predictable results. By March 2008, Amyris had formed a new partnership with Sanofi-Aventis, a leading pharmaceutical company based in France. Achieving the goal of mass-producing low-cost, microbial-based artemisinin would require greater fermentation capacity, and Sanofi could provide that capacity. Less than a year later, in February 2009, Amyris published an article announcing the achievement of their milestone: the production of 25 grams per liter of amorphadiene through *E. coli* fermentations, proof that commercially viable levels of artemisinin could be produced through Keasling's process.[9] And Keasling intends to continue to explore strategies for increasing the yield.

It's hard to imagine anything more valuable than a new medicine that would save the lives of more than 1 million children a year. But what about the new model that led to that medicine? It is the model itself, combining science, philanthropy, and market economics, that may prove to be the lasting legacy of both Keasling and Hoffman. What's more important than the fact that they may succeed in developing ways to cure or prevent malaria—as important as that is—is that they have eliminated the market gaps that so often undermine social solutions.

HITTING THE HIGH NOTES

It is no accident that both Jay Keasling and Steve Hoffman have worked closely with and been funded by an innovative organization called the Institute for OneWorld Health. Founded in 2000, it is the world's first and only nonprofit pharmaceutical. It is also a mechanism for market making. In some cases, it finds drugs whose patents have expired. In others, it gets pharmaceutical companies to donate the intellectual property from research and development efforts for drugs they've abandoned because they lack the potential for profit. Either way, OneWorld Health aims to pick up the pieces, produce the drugs at relatively low cost, and make them available to the world's poor. Since OneWorld Health doesn't need to make a profit, it can concentrate on fulfilling a mission that is purely altruistic.

The nonprofit pharmaceutical was the idea of Victoria Hale, forty-nine, who, with a Ph.D. in pharmaceutical chemistry, worked at the Food and Drug Administration's Center for Drug Evaluation and Research for four years at the beginning of her career. Shortly before Steve Hoffman left the navy to get business experience at Celera, Hale left the government to do the same at a biotechnology firm called Genentech. She used that experience to start Institute for OneWorld Health in 2000. Like both Hoffman and Keasling, she knew that the effective medical solutions were out there, but that they weren't being manufactured and distributed at the necessary scale.

In launching the Institute for OneWorld Health, Hale faced the kind of challenges inherent in reimagining something as established as the pharmaceutical industry. She recalled for the *Chronicle of Philanthropy* that it took ten months for the IRS to grant OneWorld's nonprofit status: "There was no precedent for them to comprehend the concept of a nonprofit pharmaceutical company until we offered the analogy of public versus commercial television . . . which serve different audiences, provide different products, and are funded differently."[10]

In October 2007, Hale and Keasling spoke at the annual PopTech conference in Camden, Maine, on a panel called "Changing the Paradigm." Notwithstanding the overused cliché, it was a significant and telling title. The focus was not on malaria or science but on new ways of thinking to solve seemingly unsolvable problems.

The audience did not have global health, tropical disease, or malaria eradication on their agenda. People who attend PopTech come from a wide variety of fields and are bound together only by an interest in the trends shaping the future. The chemistry and biology that dominated Keasling's and Hale's working hours were complex for the layman and almost impossible to present or summarize. But not so the expansive thinking, leaps of imagination, and collaborative partnerships they had brought to their work. Those had applicability to a wide range of issues and were relevant to the large and diverse audience.

Hale and Keasling had just flown directly from the Malaria Forum in Seattle hosted by the Bill and Melinda Gates Foundation. I watched them on a live video streaming that PopTech offers. Victoria Hale began, as I'd seen her do many times before, with a few images on slides—a typical Indian village with a high incidence of visceral leishmaniasis, a parasitic disease also known as black fever that kills hundreds of thousands a year. Hale paused on a slide of women and children, some of whom had red hair due to nutritional deficiencies. She described them as "invisible people in our world, voiceless, and many are women, which means really voiceless."

The black fever parasite is spread by the bite of a sand fly. The parasite goes into the bone marrow and suppresses white cells. The cure costs $300 but the average income in Bihar in eastern India is 30 cents a day. The Institute for OneWorld Health also works on malaria and diarrhea, currently in India, Bangladesh, and Nepal, but soon in Sudan and Brazil as well.

Hale went onto describe her own "personal journey, which began at the FDA." She said, "I was at the FDA for five years. It was a fabulous job. And then I was at Genentech and had a fabulous time, until one day I wasn't having fun anymore and resigned."

"I wanted quiet and to settle the confusion in my mind," Hale recalled. "Genentech was developing great medicines but for fewer and fewer people. Biotech products are very expensive. I felt this combination of pride in the talents of all those I worked with and shame that as an industry we weren't doing all we could. If you know that more could be done, how can you not do it?"

The problem, as she saw it, was that "only the pharmaceutical industry knows how to make new medicines. The industry had to be engaged." "So I designed an experiment to see if a pharmaceutical company could be driven not by profit but by venture philanthropy," explained Hale, referring to a type of philanthropy in which donors act like venture capitalists, investing in an enterprise's team and capacity, and establishing specific benchmarks that must be met for further investment. "We wanted to see what happened if we took out that one variable, profit. Our proof of concept, finally, was with a drug for visceral leishmaniasis that we got down to $10 a dose."

Hale is a scientist who recognizes the limitations of science. She emphasized that "the obstacles are not technological but human obstacles: lack of will, political challenges, competing priorities, not making a decision to commit. My end goal was to bring the industry back to diseases of poverty, to bring the industry back to where it once was, and to where many of the employees of the pharmaceutical industry want it to be."

In talking about those without a voice, Hale found hers. She often is asked to speak about drug development, social entrepreneurship, philanthropy, or corporate social responsibility. But more and more she brings the conversation back to "diseases of poverty," to those who have no voice, and to the choices we can make or not make to help them.

After the Q&A session, Hale closed by saying, "Technology is the easy part. The difficult part is after you have the technology. If you didn't begin with thought about who the technology should impact, it will be mismatched."

She is well on the way to proving that her idea can work: "We knew there were people who worked at pharmaceuticals who would want to help, but we had no idea there would be so many and at every level," Hale said. "Most of them got into the field to save lives and then found that many years later they were part of an effort to help a large corporation make more money. We give them an opportunity to get back to that original impulse."

As she told the journal *Nature*, "There is a growing appreciation that a lot of IP [intellectual property] exists that many people—the discoverers, the owners, the people in the world who are in need—all agree should be moved into the public domain, but there's nowhere for it to go. So, let's put it in this non-profit sector and see what happens."[11]

The OneWorld Institute's first successful drug was an antibiotic called paromomycin for visceral leishmaniasis, which won regulatory approval from the government of India on August 31, 2006. If it is the success Hale anticipates, a new industry will likely be born.

Though OneWorld Health is a nonprofit, Hale had to make decisions with the discipline of a business executive accountable for returns on investment. "We need to ask whether anyone has the will to distribute a particular drug if we developed it today," she told the PopTech audience. "For instance, and it's painful to say this, we believe that developing a drug for sleeping sickness is just not an area on which we should spend our energy right now, because the countries in which sleeping sickness is an issue are at civil war and aren't spending on public health."

Two of the relationships that Hale has built are with Keasling and Hoffman. In 2004, the Institute for OneWorld Health received a $43 million grant from the Bill and Melinda Gates Foundation to create a three-way partnership between the Institute for OneWorld Health, UC Berkeley, and Amyris Biotechnologies to significantly reduce the cost of artemisinin. She first visited Steve Hoffman when he was still at Celera, seeking advice and funding, and later she helped lay the foundation for the investment that the Gates Foundation made in Sanaria via the PATH Malaria Vaccine Initiative.

Hale told *Nature*: "Five years at the FDA taught me one thing in particular: the success of a product depends primarily on the product team. The drug and its qualities are often secondary. There are a good number of average drugs on the market that succeeded because they had ace project teams, that overcame every obstacle."[12]

When Victoria Hale delivered a keynote address at the Social Enterprise Conference at the Harvard Business School in 2007, more than four hundred students from the B-School, the Kennedy School, and other nearby colleges filled the Burden Auditorium on a Sunday afternoon, many of them having been inspired by the speakers to consider a career in public service or the nonprofit sector.

Hale spoke quietly, as if presenting the results of academic research, but it was clear from the words she chose, as well as her track record since 2000, that a passion, perhaps even an anger, fuels her work. She again used the slide presentation to show the faces of the poorest of the poor around the world, particularly in the Indian state of Bihar. "This is the twenty-first century. Why does that happen?" she asked softly, almost as if addressing herself.

Hale told the students why she started the Institute for OneWorld Health:

These people don't need the same medicines we do. There is a problem with how we make medicines. Medicines are miracles. There is an incredible beauty and power to them. In 2000 I began pulling together scattered ideas based on both my pride in the industry and my shame that we hadn't figured out how to make medicines for all. We went to India first because we knew we could have success early on. The size and scope of the success is less important than that it be a success. So as we adopted the mission of making safe, effective, and affordable drugs, we wanted to develop an organization that would show the industry a new way of working.

The increased interest in so-called "neglected diseases," fueled by international politics as well as the massive financial commitment of the Gates Foundation, gave Hale and her organization the perfect wave to ride.

Hale's political savvy and practical side came through in the talk, too:

Mainly, we need to engage the pharmaceutical industry. If we don't it is a missed opportunity and we won't be all that we can be. The industry does not know who the poorest people are and we can't expect them to know what needs to be done. But the carrot is more powerful than the stick and we must ask: What is it you need in order to engage? And also, What are you afraid of? Corporations want to do well and good and we should respect their individual capacity and potential. Don't presume you know the answers. Ask! And balance patience with persistence because it always takes longer than you hope or dream.

She concluded by urging the audience members to "trust the universe. If your intentions are good, the universe brings you what you need."[13]

"HOOKWORM HAS ONLY PETER HOTEZ"

If the development of the malaria vaccine—and the trials and tribulations of Hoffman, Keasling, and Hale—were the only story of its kind, it would still be interesting and instructive. But the nature of vaccine development for those most vulnerable and voiceless, as viewed through the long tunnel of time and shaped by the lack of markets, virtually creates an ecosystem: It's an ecosystem that serves as a breeding ground for the imaginations of unreasonable men. The life and work of Peter Hotez, who has also devoted a long career to pursuing a vaccine, though for a different disease, demonstrates the broad applicability of the lessons we've learned so far. He has worked against the same long odds, and the entrepreneurial strategies and qualities of character required for success were also the same.

Hookworm is a debilitating parasitic disease that afflicts more than half a billion of the world's poorest children with worms. These worms, using their teeth, adhere to the inside of their victims' small intestines, tearing away at blood vessels and feeding on hemorrhaging blood. You do not want hookworm.

Some 44 million pregnant women around the world become infected with hookworm annually and deliver babies of low birth weight as a result. These pregnancies are associated with higher than normal infant and maternal mortality. And that's not all, for the hookworm has no particular preference for pregnant women: Overall, 740 million people a year are afflicted with this blood-sucking intestinal parasite.[14]

Of course, none of them will be in the United States. Like schistosomiasis and lymphatic filariasi, hookworm is a neglected tropical disease whose prevalence and persistence is related as much to economic conditions as to medical conditions.

For kids, hookworm, though not typically fatal, means stunted height and weight as well as suppressed IQ and cognition. "There are periods in life when one is wormier than other periods, and peak period of worminess is age three or four through age fifteen," explained Hotez when I visited him at his office at George Washington University in September 2007. "In rural villages in Guatemala, or any rural village in Central America, for that matter, 100 percent of the kids will be infected. When you are feeding a hungry child, you are feeding the worms first." Hotez is president of the Sabin Vaccine Institute, where he founded the Human Hookworm Vaccine Initiative, and he also serves as Distinguished Research Professor and Walter G. Ross Professor as well as chair of the Department of Microbiology, Immunology, and Tropical Medicine at George Washington University.

Hotez is also the only scientist in the world who is seriously trying to develop a hookworm vaccine. Despite how widespread it is, hookworm infection has received very little attention in the scientific community, and as a result, hookworm vaccine development is not only demanding but lonely work. There is not one paper in the literature that has been written on this leading cause of growth retardation. "Some diseases have Bono or Angelina Jolie as their champions. But hookworm has only Peter Hotez," said Hotez, underscoring the obscurity under which he labors.

Malaria vaccine researchers are competing fiercely to get to market with the first effective vaccine. Approaches vary substantially in terms of the science involved. Rival researchers are polite but mostly dismissive of one another. Still, they constitute a fraternity of sorts, often reconnecting at conferences and seminars, sharing data, and debating the latest developments. Likewise for research in AIDS and many other diseases. But for Hotez, there are no rivals, and no competing vaccine development project: "No, just me," he told me when I asked.

Hotez's grey cinderblock office at George Washington University reinforces his isolation. The phone doesn't ring all that often. When we toured the lab we met only two other technicians, working alone in two small rooms down the hall.

For many years Hotez scraped by with almost no support. "Neglected diseases, neglected scientists," he said with a grin. He delved into the reasons that hookworm is neglected. "First of all, it doesn't kill. It is chronic, and debilitating and disfiguring, but many of the neglected diseases don't kill and so they don't rank high in rankings of factors that cause mortality."

Eventually the Gates Foundation came through with two multimillion-dollar grants. "I don't think they are all that interested in hookworm," Hotez confided, "but they are interested in the model we've developed of technology transfer to what we call Innovative Developing Countries (IDCs), like Brazil, China, India, and others."

In an interview for the Department of State's electronic journal (America.gov), Hotez explained:

> Product Development Partnerships . . . will actually include what we call public sector vaccine manufacturers in developing countries. . . .
>
> In Washington, D.C., we've been able to make pilot-scale amounts of vaccine for early-phase clinical testing, which is underway in Brazil. The problem is the amount we can make in our laboratories through the PDP here in Washington is limited, and certainly not enough to vaccinate all of Brazil or all of the Americas.
>
> So we've now partnered with an organization known as Instituto Butantan, which makes 86 percent of the vaccines for Brazil. . . . They're coming up here; we're going down there and transferring our technology so that they can do the scale of production for all of the Americas.[15]

As if to underscore how far he's come, he chuckled: "I even have a hookworm movie." He wasn't kidding: He turned to his computer to call up a video taken during a routine colonoscopy that included footage of a hookworm at work. We watched a five-minute excerpt of it together.

Hotez seems to have been destined for this work. As a thirteen-year-old in 1971 he had on his bedside table a volume of *Manson's Tropical Diseases*, which, through twenty-one editions since 1898, has been considered the bible of tropical medicine for both clinicians and researchers.

Upon his enrollment at Rockefeller University he was told that Rockefeller's students were "supposed to do remarkable things." He took the words to heart. When he read a famous 1962 paper by parasitologist Norman Stoll describing hookworm as "the great infection of mankind," his course was set.

In a lecture in 2006 at the University of Georgia, Hotez described the long process of making a vaccine, beginning first with dog hookworm. It required one to collect the necessary ingredient of worm spit, which, because you can't get enough of it from the worms themselves, means turning to genetic engineering. "It took twenty-five years of work to develop a viable strategy for this disease—and that was the easy part!" Hotez recalled. Hotez emphasized that there was essentially no way to create a company to manufacture the vaccine without losing money, describing the project as "the biomedical equivalent of the Broadway play, the Producers—an intentional flop—a guaranteed money-losing enterprise." But he also cited Gandhi's aphorism that "no movement ever stops for lack of funds," a telling point that revealed something of what makes Hotez tick.[16]

Despite the idealism that became evident at moments like this, Hotez is nothing if not practical. One vaccine, using a recombinant version of an enzyme, proved effective, but it was costly and difficult to manufacture in bulk, so

Hotez devoted his energy to developing a different vaccine. If his ultimate objective is to save lives, he does not have the luxury of just developing an effective vaccine for hookworm, as difficult as that is, but must develop a different vaccine, one that is effective but can be delivered for less than $1 a dose.

"If you can't make it cheaply, you might as well not make it at all," Hotez told *The Scientist*. "We have to build into our design process the ability to deliver this vaccine at less than $1 a dose."[17] This provocative assertion has obvious implications for global health science and medicine. It also suggests a different way of approaching social science. It means at least two things.

First, if you contract a disease in a developed country and your prayers are answered by a cure being discovered, you will get that cure and you will live. But if you contract a potentially fatal disease in a developing country, the discovery of a cure does not necessarily answer your prayers, because there may be no economic way to get that cure to you, even if, by not getting it, you will die.

Second, because the life and death stakes are so explicit and unambiguous when it comes to neglected tropical diseases, vaccine developers must build the economics of manufacture and distribution into their work from the beginning. Consider what is being asked of them. No one expects university economists to know how to isolate protein molecules in lab flasks to determine which ones trigger immune reactions and might be candidates for a vaccine. But doctors who are able to do just that in their labs must also master the rigors of international economics if they want to see their efforts to prevent disease come to fruition.

Businesses face this issue head-on every day. So must the nonprofit sector.

I've never heard the term "gross margin" raised in a nonprofit context, although clearly that is what Hotez is getting at when he says of his vaccine efforts: "If you can't make it cheaply, you might as well not make it at all." This is not a new principle, or a very complicated one, but neither is it widely appreciated or subscribed to in the social sector.

If one accepts that there is a moral obligation to share our strengths and intellectual gifts to develop solutions to human need, then the moral obligation may be even greater to ensure that such solutions, including vaccines and other preventive measures, are not accessible only to the privileged few.

Economic laws and market forces may be morally neutral, but our willingness and discipline to embrace and marry them to social science represents a choice. It is a choice that we have yet to make and take full advantage.

Most vaccine developers devote their entire lives to creating and testing a vaccine without ever seeing their work finished. Hotez is no different. Now at the pinnacle of his profession, he has been working on hookworm since medical school. As complex as the scientific challenges are, the economic challenges may be even greater. "The hepatitis B vaccine started at $150 a dose," Hotez told me. "It took thirty years before it penetrated the population."

But what interests me most about Hotez is not only the science of what he is doing, and the economics of it, but also the determined and sophisticated effort to build political will where it does not exist, and to do so by projecting a voice where there has been silence.

There are 540 million children, some halfway around the world and many here in our own hemisphere, whose intestines are literally crawling with blood-sucking worms. There are tens of thousands of doctors around the United States and the globe, but there is only one who devotes all of his waking hours to doing something about hookworm.

In an article in the prestigious *New England Journal of Medicine*, Hotez and some of his colleagues attempted to combat a critical notion that contributes to the plight of neglected diseases. Scientists know that more people are dying from HIV/AIDS, malaria, and diarrheal diseases than are dying from hookworm and some of the other tropical diseases, and they conclude that the more fatal illnesses must be given greater priority. As a result, considerably more talent and money go into those endeavors. But by adopting a different metric, one of "disability adjusted life years," or "DALYs," Hotez's team said, the neglected tropical diseases can be shown to constitute large burdens on the health and economic development of low-income countries. Indeed, in terms of DALYs, the neglected diseases rank closely with the better known malaria and tuberculosis. The obvious conclu-

sion: Some of that talent, and money, needs to be going toward combating them.[18]

THE BILL AND MELINDA GATES FOUNDATION

History has traumatized and scarred the malaria community. Several times over the past fifty years it was believed
that the world was on the verge of eradicating malaria. The
result of premature celebrations was a decrease in focus,
funding, and research, and soon a dramatic increase in the
prevalence of the disease. Too often the malaria parasite
has been underestimated. Discussion of eradication has
been seen as somewhere between naïve and recklessly dangerous. Melinda Gates changed all that.

On October 17, 2007, the Bill and Melinda Gates Foundation hosted a three-day forum in Seattle on malaria, bringing
in prominent experts from around the globe. Melinda gave
the opening address. Her speech was candid and coura-
geous, passionate and provocative. Historians may look back
on it as a pivotal moment. After recounting some of the history of the disease and the way it has plagued people around
the world, she said:

> We wouldn't let it happen here. We shouldn't let it happen
> anywhere.
>
> But over the course of the last century, malaria changed
> from a disease that afflicted a broad range of countries to
> a disease that affected only poor countries. It changed

from a celebrated cause of our scientists and politicians to a source of suffering that the rich world was willing to accept and the poor world was helpless to prevent. . . .

Bill and I believe that . . . advances in science and medicine, your promising research, and the rising concern of people around the world represent an historic opportunity not just to treat malaria or to control it—but to chart a long-term course to eradicate it.[19]

Anticipating the concern of those in the audience with far more expertise than her, she added:

We know that the word "eradication" is troubling to many people with deep knowledge of malaria. It's an . . . audacious goal. . . .

. . . But to aspire to anything less is just far too timid a goal for the age we're in. It's a waste of the world's talent and intelligence, and it's wrong and unfair to the people who are suffering from this disease.

The goal of eradicating malaria has the power to create great expectations, grand efforts, and record funding. When you ask people to donate time and money to save lives, they can be very generous. When you ask them to give time and money to eradicate a disease, their generosity can multiply. Those are the benefits. They are also the risks. If high energy and high expectations don't lead to success—it saps money and morale. People give up. Governments, foundations, and corporations cut their funding, malaria surges back—and gains can be quickly wiped out.[20]

Gates gave three specific reasons why we should embrace the goal of eradication. The first one was that it was the ethical thing to do. "Every life has equal worth," she stated. The second was economic: "If we plan only to control malaria, we will never eradicate it. That means we will keep bearing forever the human costs of malaria, even as we keep paying forever the financial costs of trying to treat and control it." The third was epidemiological: "Without eradication, we will continuously adapt our strategies to the parasite and the parasite will continuously adapt to us—in a back-and-forth battle that will never end."[21]

Both Melinda and Bill Gates made the case that with enough time we could develop the partnerships, political will, and scientific breakthroughs necessary to eradicate malaria. It would take relentless research, coordination, and especially long-term commitment. And eradication would require intensifying efforts as fewer and fewer people were infected, which may sound counterintuitive but is true. Political will and funding diminish when the mortality is reduced.

The pros and cons that Melinda Gates articulated were almost identical to those we'd discussed at Share Our Strength when we had been debating whether to set a goal of ending childhood hunger in America—not reducing it, but ending it. Many of the experts in our community cautioned us about the complexities of measuring our progress and the risk of failure. But as Gates understood in making her remarks, the experts are often expert in what has been, but not in what could be.

"The world is failing billions of people," Bill Gates told the World Health Assembly in Geneva in 2005:

> Rich governments are not fighting some of the world's most deadly diseases because rich countries don't have them. The private sector is not developing vaccines and medicines for these diseases, because developing countries can't buy them. And many developing countries are not doing nearly enough to improve the health of their own people.
>
> Let's be frank about this. If these epidemics were raging in the developed world, people with resources would see the suffering and insist that we stop it. But sometimes it seems that the rich world can't even see the developing world. We rarely make eye contact with the people who are suffering—so we act sometimes as if the people don't exist and the suffering isn't happening.[22]

In October 2008 I flew to Seattle to check in with the global health experts at the Gates Foundation on the progress of their malaria vaccine initiative. Joe Cerrell, director of global health and advocacy for the foundation, echoed his boss in explaining that, first, markets don't work for 2 billion of the world's poorest, and, second, "sheer visibility" is a challenge, in that "the problems are 7,000 miles away and don't factor in to the psyches of those who could make a difference." Perhaps most important, the Gates staff had come to learn, as Steve Hoffman and Peter Hotez had, that "the best science is not always the best solution," in the words of Tom Brewer, senior program officer for infectious disease at the foundation. Sometimes the best science is the science one can afford to scale.

With the stakes so high, one might expect more effective leadership and coordination among the various malaria experts and activists. But for decades there was very little leadership, because there were few resources and little to lead. Malaria's victims were too voiceless and invisible to attract funds, talent, or champions to their cause. The fight against malaria was more like a game of whack-a-mole. Each time the parasite was attacked from one direction it popped up in another. There was no one calling the shots.

By any measure, the battle was too underfunded to be considered a fair fight. A study by the Malaria R&D Alliance, an international coalition of malaria research groups, found that spending on malaria research and development in 2004 amounted to just $323 million, less than 0.3 percent of total health spending worldwide, even though malaria accounted for 3 percent of the global disease burden.[23]

Of that $323 million, $129 million came from the U.S. government. European governments contributed $31 million, and the private sector, $39 million. The nonprofit sector provided 32 percent of the funding, with an investment of $103 million.[24]

But even a few hundred million dollars gets spread too thin when it has to stretch across a planet with such large swaths of territory hospitable to mosquitoes and malaria. A 2008 report by the Kenya Medical Research Institute showed that only twenty-four countries received more than $1 per person for each person at risk, and only seven countries received more than $4 per person. The Democratic Republic of the Congo, Côte d'Ivoire, and Pakistan received only 11 cents annually per person at risk. Sixteen countries, including seven of the poorest in Africa and two of the most densely populated at-risk nations (India and Indonesia), received less than 50 cents for each person at risk.[25]

It was hardly a scenario for attracting the best and the brightest to the field, stimulating innovation or entrepreneurship, or seeding a new generation of doctors willing to commit to helping the most voiceless and vulnerable people in the world. Tropical medicine had become like a historic landmark in a once trendy neighborhood that time had passed by and that had long since fallen out of fashion. No one looked to move there anymore, and hardly a penny had been spent in ages to spruce things up. Instead, most young medical professionals seemed to be clamoring for lots in the newer subdivisions of Genomic Gardens or on Dermatology Drive.

Resources remained scarce for decades. Then one day the Bill and Melinda Gates Foundation moved in, made global health a top priority, and began awarding large and numerous financial grants based upon their conviction that something needed to be done. Everything changed. Tropical medicine and neglected diseases were once again fashionable and funded. Scientific enterprises came back to life. Biotech boutiques opened their doors. The activity attracted talent. Labs working on everything from dengue fever and malaria to schistosomiasis and leishmaniasis enjoyed a surge of postdoctoral applicants from around the world.

In 2005, the Gates Foundation announced that it would make grants of $258 million over five years to support malaria treatment and prevention. The U.S. government was virtually shamed into keeping pace. By 2007, funding for malaria control, including bed nets, improved diagnostic techniques, and drug and vaccine development, had reached $1.5 billion, with 34 percent coming from the U.S. government. Funding from international donors tripled in just three years, going from less than $250 million in 2004 to $701 million in 2007.[26]

The Presidents Malaria Initiative (PMI) is led by the U.S. Agency for International Development and is implemented jointly with the Centers for Disease Control and Prevention (CDC). PMI funding has steadily increased, going from $30 million in fiscal year (FY) 2006 to $135 million in FY 2007, $300 million in FY 2008, and $300 million in FY 2009. PMI funding for FY 2010 is expected to reach $500 million.[27]

Though funds devoted to malaria have doubled since 2003, most experts estimate that four times the current amount of funding is necessary to meet the Millennium Development goals of cutting the number of malaria cases in half by 2015. But money is just part of the story. Mark Grabowsky of the Global Fund to Fight AIDS, Tuberculosis and Malaria wrote in a commentary for *Nature* that "adequate funding for malaria control is an important first step, but unprecedented coordination, planning and operational support will be required to achieve the goals."[28]

Putting up so much money earned the Gates Foundation the right to coordinate, plan, and lead. Sometimes, such influence provokes a resentful backlash. The foundation staff is appropriately sensitive to the fine line between leading and coercing. Foundations can be notoriously heavy handed in influencing the direction and even agenda of their grant recipients, rather than just supporting them. And so lip service is given to the notion that the Gates Foundation is "only" a funder. In a speech in the fall of 2004 at the University of Washington, Bill Gates jokingly referred to the awkward dynamic between grant maker and grant recipients, saying that success in the philanthropic field was much harder to measure than success in his other endeavors. "In business, the market tells you when you've failed. In science, your instruments tell you," Gates said. "In philanthropy, no one tells you. Everyone wants to be your friend."[29]

But the Gates Foundation leads by default as well as by intent. It's still early to say whether its leadership will be embraced or resisted, but at least the opportunity for leadership has been created. The toughest tests still lie ahead.

For now, there have been many beneficiaries, not least of whom is Steve Hoffman, who with Gates funding was able to move his tiny lab from a small strip mall in Rockville to the sprawling manufacturing facility that helped qualify his vaccine for clinical trials.

CHAPTER 8

SANARIA HAS ITS
DAY IN THE SUN

The multimillion-dollar effort to eradicate one of the world's deadliest dis-
eases received a significant but controversial boost yesterday when scientists
announced the creation of genetically modified mosquitoes that cannot pass
on malaria. . . .

The strategy is likely to prove contentious as it would require the un-
precedented release of tens of thousands of GM organisms into the wild.
But it has raised hopes among scientists, some of whom believe it may be
powerful enough to finally bring under control a disease which strikes 300
million people a year and causes more than 1 million deaths, mostly of
children in sub-Saharan Africa.

—Ian Sample, "Malaria: GM Mosquitoes Offer New
Hope for Millions," *The Guardian*, March 20, 2007

D R. PEDRO ALONSO, FROM THE Hospital Clinic of the University of Barcelona, flew to the October 2007 Gates conference in Seattle, and in front of a large audience of experts,
which included Steve Hoffman, announced the release of a

report in *The Lancet* showing that the RTS,S malaria vaccine candidate had proven to be more promising than anyone had expected. International health workers had administered the vaccine along with other routine immunizations to 214 babies in Mozambique in 2007, and the report was claiming that it had cut the rate of malaria infections in the infants by 65 percent. "It is hard to overstate what a major step forward this is," said Rip Ballou.[1] It garnered headlines in the *New York Times* and worldwide, but the news was quickly overtaken by the inspiring yet controversial challenge from Bill and Melinda Gates at the malaria forum in Seattle.

Fast on the heels of that news, with some of those who'd been at the Gates event in Seattle heading east via red-eye flights, came a day-long ceremony celebrating the opening of Steve Hoffman's new facility for the clinical manufacturing of a live, attenuated, sporozoite malaria vaccine, the only one of its kind in the world. It was yet another advance— perhaps the most unlikely of all—in the battle against the world's most lethal infectious disease for children. (The media coverage itself was revealing—*The Times of India*, Reuters South Africa, and Australia's News-Medical.Net all reported it, but not the Washington, New York, Chicago, or Los Angeles papers.)

Steve Hoffman's ability to get the most eminent members of the malaria world to the opening of a building long before a viable vaccine had been produced within it was a tribute to his ingenuity and competitiveness. Without sufficient current results that could match RTS,S, Hoffman was selling the promise of the future, and this new facility embodied it.

The opportunity to invite Bill and Melinda Gates to attend Sanaria's ribbon cutting had been too great for Hoffman to resist when he was with them in Seattle. With hundreds in attendance at that conference, Hoffman posed a challenging question during the Q&A after Bill Gates spoke. When the chance to ask questions was announced, Hoffman seized it in order to put his fledgling venture on the map.

"So I ask this pointed question," recalled Hoffman, telling me about the incident later. "Given the fact that the HIV virus has nine genes and the malaria virus has 5,300 genes, when is their funding going to equal the complexity of the challenge? Gates danced around it just fine. But then at the coffee break, someone comes over to me and says 'Are you Dr. Hoffman?' And I say yes. And they say 'I'm sorry, there's been a mistake. You're supposed to be sitting at Mr. Gates's table!'"

"So I move over to an empty seat at Gates's table," Hoffman told me, "and there I am, on GSK's big day, the day they make this big announcement, and I'm sitting during the rest of the session with Bill Gates." Hoffman asked Regina Rabinovich, director of infectious diseases for the Gates Foundation, about inviting the Gateses to the opening, and she said, "We don't go to building openings. It's premature. The vaccine hasn't been proven yet. We'd rather wait to celebrate the results of clinical trials." Hoffman thought, "I know it's just a building opening, so I realized that celebrating would be a little controversial, but I have to keep my people going. So I invited them anyhow."

Hoffman sent his staff an e-mail that day. It read:

> Dear Sanarians, I am at the Malaria Forum sponsored by the Bill and Melinda Gates Foundation. There are 250 individuals here from all over the world who are the leaders in the effort to control malaria. These include 4 Ministers of Health, the Director of UNICEF, the Director of NIAID, a British Member of Parliament, etc., etc. Yesterday morning the session started with talks by Bill and Melinda. They were inspiring and enunciated the challenge to everyone in the room that the goal was to eradicate malaria not just control malaria. Few in the room had been willing to embrace this concept even the day before when there was a heated discussion about setting goals beyond control. As you know our website has stated the following for the last 6 months, "Malaria Eradication Through Vaccination."

The e-mail continued in the same vein, rallying the troops with new causes for optimism and new reasons to believe that their work would ultimately be vindicated. The truth was more complex. Hoffman had his new space, and he had sat at the Gates table. But he knew full well that was not enough: "[Sanaria had] twenty months' worth of cash" before it would go negative, he told me.

Securing grant money is an endless process and is endlessly frustrating. Sanaria's expansion had been possible thanks almost entirely to the grant of $29.3 million of Gates's money through PATH Malaria Vaccine Initiative (MVI), but even this grant had once promised to be as much as $80 million. Moreover, grant givers can be very interfering in the eyes of visionary entrepreneurs: "Most of them have never done clinical trials. Yet they are telling me how to design mine," was how it seemed to Hoffman.

Nonetheless, the money allowed Hoffman, by August 2007, to complete his state-of-the-art vaccine manufacturing facility in Rockville, Maryland, but progress was not always smooth. For one thing, the MVI wanted to resolve whether Sanaria could continue to contract with Hoffman's wife's company, which was called Protein Potential. His wife, Kim Lee, served as vice president of Sanaria, and Steve served as chairman of her firm. The MVI people were concerned about a conflict of interest—or, more likely, the appearance of it. "So I've had to spend thousands of dollars on lawyers," Hoffman told me, in the annoyed tone of one who has experienced the legal profession exclusively as a nuisance.

Sanaria actually had no choice but to work with Protein Potential. No other company could do what Kim Lee's could to sterilize and purify the vaccine they had developed. It was a highly specialized technology and perhaps the key to overcoming the regulatory obstacles that the rest of the vaccine community assumed would doom Hoffman's chances. "I'm astonished that we've been able to do this and keep it sterile," Hoffman said. "I really find that astonishing. . . . That's all Kim Lee. Her standards are very exacting." Still, they had to ensure an appropriate way for the transactions to be conducted through a third party.

It had only been a few months earlier, in March 2007, that I'd visited Hoffman on what he declared to be a "banner day." On each of the previous occasions I'd visited Sanaria, he had also declared it to be "banner day." This was either luck, coincidence, or an indication of Steve's gift at a form of self-promotion that manages to get others to be as invested in and enthused about his work as he is. It may also be an internal motivation strategy. The road to a vaccine is so long and problematic, such a long-odds gamble, that those who succeed are those who find ways to keep going across an inhospitable desert, squeezing enough sustenance out of each small victory and milestone, while those around them drop behind or perish.

"Last week we actually manufactured the vaccine," he told me that March. "Right here. Three hundred vials of it. And we had them tested to see if they were safe, sterile, stable. The tests came back today." Then, speaking slowly for emphasis, he added, "They are safe, sterile, and stable. This is what we need for the preclinical toxicology studies, which will be with rabbits. When those studies come back, we file an Investigational New Drug Application and the FDA has to get back to us within thirty days. Then we go to clinical trials and within six months of putting it into the arms of volunteers, we will know how effective it is."

Steve suggested we ride together to the building that would house the new lab. We got in his nine-year-old Jaguar to drive out to Rockville's Biotech Corridor, just fifteen minutes away. On the highway, to my surprise, the car jolted and sputtered, and Steve visibly struggled to keep it moving, pumping the gas pedal, gripping the jerking steering wheel with both hands. It was not what I had expected from a polished and shiny Jaguar. It seemed just a matter of time before we would be stranded along the side of the road. But Steve never made mention of it. He kept up his monologue about a BBC film that had been made about malaria vaccines, as if I thereby might not notice, as if the only thing that could bring the vehicle to a halt was acknowledging its shortcomings. It was not a bad metaphor for Sanaria's journey so far—an impressive-looking operation that seemed at times to be lurching from start to stutter to almost stop, with a driver who kept his eyes straight ahead, powered by confidence, faith, or the simple conviction that there was nowhere else he'd rather be going.

Sanaria would have several floors of the building that also housed a few other companies, but it seemed all but abandoned as we toured it. We never saw another human being. This space was previously a lab for another biotech company, and the lab benches and tables were still there, waiting to be torn out. It was at least ten times the space he'd had to work with hitherto, with room for eight dissectors. There were individual offices and large, enclosed areas for manufacturing, including the autoclave, which would steril-ize equipment via high-pressured steam; the 14,000-pound irradiation machine; and the insectary. The life sciences supply company Fisher Scientific would store the finished product. It was a professionally designed facility, equipped for the purpose: "The lawyers and advisers really didn't want us making the vaccine at the other place," Hoffman said. "They just didn't think it would look right given the way that stuff was all thrown together." He would later admit about Sanaria's first facility: "We couldn't control the temperature. We couldn't control the air. We had floods. One time during one of those manufacturing campaigns it got to be 104 degrees outside and the air conditioner froze to the roof. So Bob Thompson, our director of operations, was up there with a blowtorch trying to unfreeze it."

There was a rags-to-riches, born-in-a-log-cabin aspect to Steve's experience that he seemed to have already concluded would be an important part of Sanaria's legend. He relished the makeshift quality of Sanaria's original lab, how little money they'd had when they'd started, the image of sitting at the kitchen table with his son, writing the business plan together. The vaccine and the storyline were both being developed by Hoffman, and in tandem.

They eventually moved into the new facility in August 2007. Steve planned to do some trial production runs after that to get the kinks out and start manufacturing in January 2008. Once the trials went well and the process was approved, Hoffman hoped to build an identical plant somewhere in Asia, probably India, and then a plant five times as large in Africa.

It was a breakneck pace, without a lot of time to spare. Nevertheless, Steve expected to keep Sanaria in the limelight as well. He had agreed to speak at a forum at Harvard's School of Public Health that would also include Jay Keasling and a representative from GlaxoSmithKline to talk about RTS,S. Steve also wanted to take a team of six from Sanaria to London for another such forum in September, this one hosted by the Royal Society of Medicine. I asked him how much he told audiences like that about how Sanaria managed to do what it did. "I don't tell them anything," Steve said. "I simply explain the results that we are getting."

The tests in Africa were to be in Ghana and Mozambique. Steve had worked in Ghana before, and both countries had established vaccination sites. "In fact," he said, "Pedro Alonso, who is a friend, told me that he will do the trial, because even though he is doing RTS,S, he is not convinced it is going to be as effective in the long term as Sanaria's. The RTS,S folks wrote a paper whose purpose was to communicate that our strategy wouldn't work. Imagine doing that. I really don't think of myself as competitive. As a scientist you want to see something work even if it is something developed by someone else. I mean you always have delusions of grandeur and you'd like it to be yours, but . . . " He left the sentence unfinished, and began again: "Ours is the only vaccine whose purpose is to prevent infection. The other vaccines slow the spread of infection in the body."

Hoffman had been thinking, too, about how this vaccine would actually get into the arms of African children if it made it through clinical trials. He was envisaging a distribution system that would bring business to Africa. And instead of relying on some government program or nongovernmental organization, he would create it himself: "Usually something like that is done through the World Health Organization, but I think by using liquid nitrogen, we can keep the vaccine potent for at least two months and have a roundhouse operation that sends it out. It could create jobs." Like so many other aspects of Hoffman's venture, this had never been done before.

I asked Hoffman about a new report that Marcelo Jacobs-Lorena and his colleagues had just published about the creation of genetically modified mosquitoes that could not pass on malaria. The strategy would mean releasing thousands of genetically modified malaria-resistant mosquitoes into the wild on the conviction that they would breed and become dominant, and therefore no longer transmit the parasite. The report had received front-page coverage in *The Guardian*.[2] Hoffman was skeptical. "The concept is such a stretch," he said. "But the media takes it and reprints it at face value without asking any questions. I'll tell you what is going on there. Donors put $200 million into the Hopkins School of Public Health and another $100 million into malaria there. So they have to have something to show them." He concedes that "Marcelo is a lovely guy." But he explained that "they are creating transgenic mosquitoes to have fewer parasites. I'm doing just the opposite. I'm creating transgenic mosquitoes to have more. Initially they came to us to get their parasites. We were growing them and they weren't."

I asked if he planned on writing any more papers. "I haven't written a paper since 2002," he responded, "and I'd written 350 in the ten years before that. But it's not necessary now to get the money."

Steve was still competitive, but he was no longer feeling unappreciated. There was more of a bounce in his step, and less of a grudge in his voice. The grants were enabling him to pursue his dream, and they were also a form of validation. Like an athlete who had never lost confidence in his own abilities, but still celebrated when signed by a major league team, Hoffman was now inside the clubhouse. He was being invited to speak at international conferences, and he was being listened to. It was a very different time from the days when Diane Griffin, the head of the Johns Hopkins Malaria Research Institute, had giggled about "that crazy thing Steve Hoffman is doing."

The feasibility of Hoffman's vaccine hinged on being able to obtain sufficient quantities of sporozoites from a process once thought to be so tedious, laborious, and fraught with technical complications as to be entirely unrealistic. Insects taken from the field have a few thousand sporozoites. Labs that raise mosquitoes from eggs to larvae to pupae to adults and feed them on parasite-infected blood get about 15,000 to 20,000 sporozoites per mosquito. Hoffman had figured out how to get more than 70,000 or even 80,000. His vaccine was the end product of a rigorous set of proprietary processes and procedures specifically designed to overcome those challenges.

Hoffman had recently returned from a Keystone Symposium in Alpbach, Austria, where he'd given the opening address to a three-day conference on all aspects of malaria, including vaccine development. On the way, he'd stopped in Geneva to make the case to the World Health Organization about recommending a standard for measuring the effectiveness of vaccines. GlaxoSmithKline was using a "time to event" analysis rather than a proportional analysis for measuring the efficacy of RTS,S. What they were really measuring was how long their vaccine delayed vaccinated participants from getting malaria, not the degree to which it prevented infection. One of Hoffman's issues with RTS,S had always been how the results of the clinical trials were reported.

Many public health officials preferred the proportional analysis, which was based on the proportion of vaccinated participants versus "control" participants who did not become infected with the disease during a specifically defined period. So, for example, in the children in Mozambique from ages one to four, RTS,S reported vaccine efficacy was 45 percent for delaying a first infection over the first six months. With the proportional analysis approach for the same data, the vaccine efficacy against acquiring an infection by age six months was 11 percent.[3]

This was a critical issue in Hoffman's David and Goliath competition with GlaxoSmithKline, which hopes to manufacture RTS,S as the world's first commercial malaria vaccine. Steve recounted for me the flaws in the process with an air of exasperation fueled by moral outrage, and, despite his assertion to the contrary, a fierce competitive nature. "Pedro Alonso is very charming and articulate and a wonderful guy, and when he talks about the results of their trials everyone nods and goes along. But anyone who really knows anything about vaccines knows that RTS,S is not preventing malaria."

Judith Epstein, a commander in the navy's vaccine research division who would eventually be responsible for conducting the clinical trials of the Sanaria vaccine, had written an article in *The Lancet* critical of Alonso's methodology. It was pointedly entitled, "What Will a Partly Protective Malaria Vaccine Mean to the Mothers in Africa?" In it she concluded:

> For a vaccine like RTS,S, which delays but does not necessarily prevent infection or clinical malaria, time-to-event analysis is needed. Nevertheless, because we are not certain whether time-to-event or proportional analysis will ultimately predict the long-term effect of the vaccine and because of the complexities of assessing a partly protective vaccine, I believe that it would be helpful for the authors to state the results of both types of analyses in future publications.[4]

What Epstein seemed to be saying was that a vaccine that delayed infection but didn't prevent it would be good, but it would not be good enough.

A GATHERING OF MALARIA ROYALTY

For the Sanaria ribbon-cutting ceremony on October 26, 2007, a large meeting room was packed with several hundred people. They represented the wide diversity of interests and opinions that have always existed, sometimes divisively so, when it comes to solving the plague of malaria. There were American scientists and European doctors, African diplomats and U.S. military personnel, the latter in dress uniforms bedecked with medals and ribbons. Some were advocates of vaccines. Others had devoted careers to bed nets, insecticides, antimalarial drugs, or the creation of public health infrastructures. Not even the vaccine advocates were all convinced that Hoffman's vaccine was the way to go.

It was politically astute of Hoffman to showcase icons from previous eras and to place his own work in the context of all the other efforts, as if to soften the break that his effort represented with what had come before. By subtly but visibly putting Sanaria on a continuum, he took the edge off of just how audacious his idea was—without diluting the audacity itself.

The attendance of some amounted to a hedging of bets. Past the midpoint of careers that careened between glimpses of glory and repeated near misses, and after decades of experiments, trials, travel, hearings, field clinics, and late-night edits of papers for scientific journals, who could take the chance of not being in the room when something as potentially historic as the first lab and strategic plan for manufacturing a live, attenuated malaria vaccine was unveiled, even if suspecting the odds of ultimate success were slim? Tropical medicine in general, and vaccine development in particular, made gamblers of doctors, researchers, and grantmakers.

The day's agenda included the obligatory elected officials, from U.S. Senator Benjamin Cardin down to the local congressman and county executive. Their interest in Sanaria seemed more parochially confined to the impact of its growth on Montgomery County's employment statistics, and to making sure their constituents knew of their role in securing the initial Small Business Innovation and Research grants that started Sanaria on its way.

But the malaria community's royalty was also attendant. Dr. Regina Rabinovich from the Gates Foundation represented the crown. At the lunch break, suitors circled around her, hoping for a brief audience. The princes included Dr. Anthony Fauci, the director of the Institute of Allergies and Infectious Diseases at the National Institutes of Health (NIH); Dr. Christian Loucq from the PATH Malaria Vaccine Initiative; and Dr. Louis Miller, the chief of the Malaria Vaccine Development Branch of NIH, who began working on malaria in 1965, eight presidential administrations ago.

If there was a queen mum it was Ruth Nussenzweig, who back in 1967 first demonstrated the effectiveness in mice of the vaccine that Hoffman has overcome technical challenges to manufacture. Stooped by age, their hair gray, and holding hands, Ruth and Victor Nussenzweig wore name tags that were unnecessary for all but a handful in attendance. I asked her if she had ever anticipated this day would come. "Never," she said quietly but emphatically, in her Austrian accent. "I never did." Throughout the remainder of the day, virtually every speaker made a point of paying homage to her work, "not so much for anything she's done lately," as science writer Merrill Goozner later put it in his blog, "but for having carried the torch through the long, dark night."[5]

To demonstrate the fragility of the chain of discovery, in which happenstance is everything, Ruth Nussenzweig told me how she had come to make her contribution. She had met Victor, her husband and collaborator, at the entrance exams for medical school in Brazil about sixty years ago. She was originally from Austria:

> Both my parents were physicians but we were chased from Vienna because we were Jews.
>
> My initial research was on *Trypanosoma cruzi* [a parasite that causes another neglected disease, known as Chagas' disease for the doctor who discovered it in 1909]. We found that if we added gentian violet—a dye—to the blood, the *cruzi* was killed. This is my *cruzi* story.
>
> We came to the U.S., went back to Brazil and then returned here. They needed an immunoparasitologist in the Department of Preventive Medicine. I became an assistant professor, then associate, then full professor, and then the first woman chair of a department at the medical school, and for many years I was the only woman chair.

I started to work on malaria in mice. For a long time it was just me and no one else. The dogma at the time was that malaria did not induce immunity, so that there could be no vaccine. This is not true. There is partial protection. People became less ill as they became older.

When I asked her if anyone had tried this before, she said:

As always, someone preceded me, it was years before, with bird malaria. Birds had been exposed to ultraviolet light. I tried irradiation. I played around with it. It needed a high degree of radiation. Intravenously with mice because you can do that with mice. You couldn't miss the result. There were no parasites in the blood.

Years later I immunized monkeys. There was a good deal of protection but not full protection. I was called to Walter Reed by Phil Russell.[6] He asked: "How would you go about immunizing humans?"

Then something funny happened. I couldn't participate in the study on humans because they were done on volunteers in prison and women were not allowed in. So the study was published by a man, a colleague right here. I was naïve. I accepted this.

Though it was other researchers who documented that the same concept would work with humans, it has always been Ruth Nussenzweig who was credited with proving that the irradiated parasites could act as an immunity-triggering vaccine. She and Steve Hoffman would see each other at conferences. In 2008 her life's work was recognized with the award of the Albert B. Sabin Gold Medal for her pioneering efforts. Whichever of the many competing malaria vaccines eventually succeeds, it will owe its genesis to her.

So it was symbolic and right that Dr. Nussenzweig should have guided the audience through a PowerPoint presentation of the vaccine's history, offering an occasional touch of nostalgia when asking, "Steve, you remember this?" "They say you cannot do better than nature. Well, one can, and Steve is succeeding," she said, almost echoing Jay Keasling's words. When she was finished she was given the day's only standing ovation.

Steve Hoffman wore a dark suit and blue striped tie, and the relaxed glow of someone whose day had come at last. He served as host, master of ceremonies, audiovisual technician, speaker, and object of admiration and acclaim. He introduced Anthony Fauci as "the most respected physician scientist in the world," and Fauci, winner of the Presidential Medal of Freedom for his work on HIV/AIDS, returned the compliment, acknowledging it as "an historic day" and explaining that he'd first met Steve in the late 1980s and had found him to be "a doer of the highest magnitude." He recalled sitting next to him at a dinner when they agreed it would take a Manhattan Project approach to address malaria. He said Hoffman had "the insight, talent, energy and drive to take a vision and follow it."

For the past few years he and Hoffman had met every six months. Praising the vaccine for being built on a well-founded scientific concept, Fauci said that "the big hurdle to overcome is the technical challenges, whether it could be produced in sufficient quantities and meet regulatory hurdles."

When Regina Rabinovich of the Gates Foundation spoke after Fauci, she explained that the foundation had searched for the infectious diseases whose toll represented the greatest inequality and the greatest disparity. She shared the foundation's introspection on the issues of "'Are we being ambitious enough?' 'What do we want to achieve 20 to 30 years from now?'" and said, "The answer to those questions probably means having to develop different tools."

Jeff Sachs, director of the Earth Institute at Columbia University and advocate for the Millennium Development goals to fight poverty in the developing world, was complimentary of Hoffman but argued that "bed nets are the tool we have now" and said he couldn't believe "how hard it is to get rich people to give a small amount of money for bed nets." Kent Campbell, the program director for the PATH Malaria Control and Evaluation Partnership in Africa (MACEPA), made the point that it was possible to actually change the epidemiology of malaria by going rapidly instead of incrementally.

Hoffman used his time at the podium to share the credit with his team, and to acknowledge the role of Ruth Nussenzweig and an early coauthor named Tom Luke, but principally to emphasize that what everyone had thought was impossible had instead turned out to be possible because of bioengineering and applied biology. Radiating infected mosquitoes weakens the parasite that resides in the mosquito's salivary glands enough that the attenuated parasite will trigger natural immunity without actually making one sick. The challenge is in then dissecting the salivary glands and harvesting enough of the sporozoites to equal 1,000 bites of a mosquito, which is nature's threshold for conferring immunity over time.

He gave the vaccine battle plan's time line, explaining that three days after it is in the liver the parasite is 3 to 5 microns in size, a micron being one-millionth of a meter. But if they have been irradiated, they stop growing, which advances the goal of ensuring that no parasite gets out of the liver and into the bloodstream.

Aseptic mosquitoes go from eggs to larvae to pupae in nine days in a flask. By day fourteen they are ready to feed on infected blood. After feeding, they sit for sixteen more days before being irradiated, dissected, and put in a vile.

The real challenge was to make sure the vaccine was free of pathogens, free of bacteria and fungi, and free of salivary-gland material (from which people can have allergic reactions). For this, Steve and everyone else acknowledge the indispensable role of Steve's wife, Kim Lee. She never spoke or presented from the podium, perhaps because her contribution represents such proprietary intellectual capital that very little could be said about it.

Hoffman was transparent about the difficulties that lay ahead for introducing the vaccine into human beings. "We still don't know how to give it. Before it was given by the bite of an infected mosquito. We can't do that. We don't know the number of doses or the interval between them, and so we have to do a dose escalation study. I don't know if this will be administered subcutaneously, or muscular, or what. There's no way to know." What is the best route? The best volume? The best site on the body? Steve is the first to admit that he doesn't know.

"THE LIVER IS WHERE THE WAR WILL BE WON"

By the end of the day at Sanaria's ribbon-cutting ceremony, the government officials had moved on to search for new limelights in which to bask and the audience had thinned somewhat, leaving mostly longtime and hard-core malaria advocates spread out among the folding chairs that stretched to the back of the room.

Superlatives continued to echo through the air. Almost everything that could be said about the path to Steve Hoffman's vaccine development strategy and its wisdom had already been said. But the closing speaker, John McNeil, from PATH, following at least a dozen other distinguished dignitaries, gave the day its exclamation point. He quoted a memorable line from Maurice Hilleman, the creator of vaccines for mumps, measles, rubella, chickenpox, and hepatitis B, and discoverer of more than three dozen vaccines in all, who is widely believed to be the most successful vaccinologist in history.

McNeil reminisced for the audience about the time Hilleman told him, "If you want to get rid of malaria, kill it in the liver. That's where the war will be won!" Since Hilleman died in 2005, he has become an even greater icon in global health circles. Quoting his recommendation for a vaccine that attacks the *Plasmodium falciparum* parasite in the liver was like quoting General Dwight D. Eisenhower on how to repel an invasion of Europe.

I thought back to the time many months before when Hoffman had leaned over the large illustration of the parasite's life cycle. Vaccinologists often fall back on military terminology. They use the language of invasion and resistance, attack and defend, kill and protect. One reason may be that many of the men and women working in tropical medicine had little choice but to enlist in the military, which used to have the best, if not the only, research labs. Their discipline, precision, and language was shaped in that crucible. They respect the parasite as an elusive and deadly adversary, one that has historically been able to withstand everything they threw at it and come back stronger.

But the effort to eradicate malaria resembles war in many ways that go beyond the language of metaphor. First and foremost, it is truly a battle to the death. The victors will survive and the losers will perish. The effort continues until, like war, the opponent is vanquished. But it is a war being fought not against the backdrop of a visible political agenda, whether extremist, ethnic, or imperialist, but instead in the nearly invisible vortex of evolution's long and seemingly infinite forces. If the Cold War was, in the words of President John F. Kennedy, "a long twilight struggle," the war against malaria is one of the longest of all evolutionary battles. It has entailed deception, reconnaissance, aggression, stealth, and lethal weapons.

Allied medical experts in every corner of the globe find themselves both competing and collaborating to identify the malaria parasite's point of greatest vulnerability. Thus, in Zambia, at the Malaria Institute of Macha, 80 miles from Lusaka, the staff of the Johns Hopkins School of Public Health study why some kids get severe malaria and some do not.

In Uganda, Sweden's Karolinski Institute and Makerere University study how the malaria parasite conceals itself in the placenta of pregnant women, causing women with their first pregnancy to lose the immunity usually found in African adults.

Tony Holder at the National Institute for Medical Research in London targets the blood stage to see what must be done to lock out merozoites from red blood cells. David Kaslow at NIH is working on a vaccine that mosquitoes would suck up in human blood. It would stop the parasite from reproducing within the mosquito and put an end to mosquitoes being able to transmit the parasite.

The Centre for Novel Agricultural Products at the University of York is using $13.6 million from the Gates Foundation to fast-track a breeding program for the sweet wormwood plant (*Artemisinin annua*) to create nongenetically modified varieties that will increase yields.

In 2010, the United Nations Children's Fund, UNICEF, will distribute 25 million insecticide-treated bed nets to protect children from being bitten at night by malaria-infected mosquitoes.

In research funded by the Wellcome Trust and the National Institutes of Health, a study led by Dr. Nick Beare of the St. Paul's Eye Unit in Liverpool has shown that changes to the retina were the only clinical sign or laboratory test that could distinguish between patients who actually died from cerebral malaria and those with another cause of death. These changes are visible with just an ophthalmoscope, making it possible to track the disease in Africa, where there's a shortage of expensive medical equipment.[7]

Malaria cases soared in the KwaZulu Natal province of South Africa after it stopped using DDT in 1996, and the reintroduction of DDT in 2000 brought the disease back under control. That is enough for other countries, such as Uganda and Kenya, to examine whether DDT could also work for them. The Ugandan minister of health, Jim Muhwezi, recently defended the plan to use DDT for indoor spraying in his country, emphasizing the need for a proactive rather than reactive strategy against malaria.

Researchers at the Johns Hopkins Malaria Research Institute, whose malaria control strategy is premised on replacing malaria-carrying mosquitoes with malaria-resistant mosquitoes, determined that genetically modified mosquitoes, known as "transgenic," fared better than their natural counterparts when fed malaria-infected blood. Such mosquitoes block transmission of the deadly parasite. The success of their theory depends on transgenic mosquitoes producing more offspring and having lower mortality than natural wild mosquitoes.

These are only a handful of the efforts underway at universities, schools of medicine and public health, field clinics, military labs, pharmaceuticals, nonprofits, and global health organizations. As in war, there are soldiers and there are generals, and among the generals a few of the most daring become known throughout their field and then beyond it.

The ongoing size of the struggle was confirmed by the final speaker at Hoffman's ribbon-cutting event. Dr. Adel Mahmoud, president of Merck Vaccines from 1999 to 2005, who now teaches at Princeton University's Woodrow Wilson School of Public and International Affairs, offered a sober note of warning: "A vaccine in the next year or two? Come on, folks, let's be real. It's going to take a lot longer than that. Eradication? It's a nice goal, but we have eradicated only one disease in all of history, and that of course is smallpox. In not one vaccine today do we understand the mechanism of protection. Not one."

The bet laid down in Rockville is that, long odds notwithstanding, man's creative genius—at least one man's—will win out in the end. In a way, it is as if Steve Hoffman has borrowed a lesson from one of Barcelona's most celebrated creative geniuses—not his friend and competitor Pedro Alonso, but the late Antonio Gaudi. The fabled, iconic architect built his models from nature, capturing the geometry of structure and adopting essential principles from the lines, shapes, fibers, and textures he observed. Believing that natural structures had millions of years of perfect functioning to their credit, Gaudi often mimicked them in his work. From seashells to beehives, from mushrooms to ears of corn, the ingenuity of nature, for Gaudi, offered clues to effective function. He once said that nature was "the Great Book, always open, that we should force ourselves to read."[8]

Steve Hoffman has done no less. Sanaria's live, attenuated sporozoite vaccine comes directly from nature's book, and by meticulously recreating the immunity that is triggered upon the accumulation of 1,000 mosquito bites, it may mirror nature in its purest form.

After three decades of experimenting with recombinant DNA technology, which is engineered through the combination of one or more DNA strands that would not normally occur together, Hoffman is pioneering a return to first principles, and especially to the kind of natural solutions Gaudi sought out.

Paradoxically, nature's purity is so complex that recreating its essence requires an engineering technology of mind-boggling sophistication. That's what Sanaria has invented. Using biomolecular jujitsu, Hoffman has managed to turn the parasite's strengths against itself. As if adopting Gaudi's famous phrase that "originality is returning to the origin,"

Hoffman returned to the origins of the Nussenzweigs' work of an earlier generation. If it works, it will be the salvation of generations to come. The applause heard within the space of just a few days—from Seattle for Melinda Gates to Maryland for Steve Hoffman—was the sound of history accelerating.

CHAPTER 9

ECOSYSTEM OF MARKET FAILURES

What could the sex lives of algae have to do with finding a vaccine for malaria and other parasitic diseases? Quite a lot, it seems, because pond algae and the mosquito-borne Plasmodium *parasite that causes malaria turn out to use the same protein to fuse their male and female gametes during sexual reproduction.*

William Snell at the University of Texas Southwestern Medical Center in Dallas and colleagues discovered that . . . when [the protein] HAP2 was knocked out in Plasmodium, *mosquitoes failed to spread malaria between mice. . . .*

A vaccine designed to block HAP2 could break Plasmodium's *reproductive cycle in people infected with malaria and so prevent its transmission to others.*

—"Algal Sex Could Spawn Malaria Vaccine," *New Scientist*, April 20, 2008

O F THE WIDE VARIETY OF approaches being tried to eradicate malaria, some are mainstream, many not, but almost all fall into one of three categories: vaccination of those who have not had malaria, medication for those who do have it, or prevention through bed nets, spraying, and the like.

One reason there are so many approaches is that the problem has been so intractable. Another reason is that there is no natural market for the solutions that would serve to control them. Markets not only deliver capital and resources, they do so in ways that are efficient. Foundations, which are insulated from market forces, do not. Most nonprofits were established in response to a market failure. Whether providing health care to new immigrants, food to the hungry, or teachers and principals to inner-city schools, nonprofits fill gaps left by either the marketplace or the government because economic or political incentives to fill them were lacking. The typical fate of many good ideas in the nonprofit sector is initial popular support that fuels short-term growth, followed by gradual, long-term growth stymied by lack of sustainable resources.

But changes are afoot in the nonprofit world, and it is no longer business (or philanthropy) as usual. Advocates for the use of market mechanisms to accomplish social objectives have grown in number and now include Bill Gates. In a 2007 Harvard commencement address that called for "a more creative capitalism," Gates proposed that we "stretch the reach of market forces so that more people can make a profit, or at least make a living, serving people who are suffering from the worst inequities."[1]

Gates did not explain what he meant by "stretch the reach of market forces." He never mentioned "social entrepreneurship." But some of the Gates Foundation grants and grant partners are illustrative of the concept and can give us a glimpse of a future for nonprofits that harnesses market forces and thereby allows them to achieve more than ever before.

A prime example is the foundation's investment in GAVI (formerly known as the Global Alliance for Vaccines and Immunization), which develops vaccines for HIV, malaria, and tuberculosis and improves access to these vaccines for children in poor countries. Gates and GAVI helped to create a new financing mechanism by which foundations and industrialized nations promise to subsidize the future purchase of a vaccine not yet available (up to a predetermined price). Called Advance Market Commitment, the mechanism is designed to encourage potential suppliers to invest in the R&D and production capacity needed to develop and manufacture the vaccines. They will be more likely to make that investment, in other words, if they know there will be a viable market for the end product. It is philanthropic spending intentionally designed to leverage commercial investment.

Gates's strategy with GAVI recognizes that the qualities that make for visionary and intrepid scientists are anathema to marketeers. Great scientists are willing to face down hundreds of years of failure in pursuit of the ultimate game-changing success, whereas marketeers, able to perceive the appetites and enthusiasms of the general public today, are less interested in the ill-defined appetites and enthusiasms of an unknown future. Scientists capable of making breakthrough discoveries are often idealists; marketeers are usually pragmatists. The scientists behind breakthrough discoveries are visionaries; marketeers look at the bottom line. Gates realized that he needed to support, even create, a market mechanism for the visionaries, whose general detachment from the mass market made them indifferent, if not ignorant, to many practical considerations.

New Yorker writer James Surowiecki described the impact and the rationale for such an approach:

> A couple of weeks ago, Gordon Brown, Britain's Chancellor of the Exchequer, made a promise. The United Kingdom, he said, would buy up to three hundred million doses of a new malaria vaccine for the developing world. . . . It was . . . a dramatic innovation in the way those diseases are fought. . . .
>
> What Brown's announcement guarantees is that if an effective vaccine emerges there will be someone to buy it at a fair price.
>
> Usually, a company that invents something useful doesn't have much trouble selling it. But vaccines—especially for diseases in the developing world—are notorious exceptions to this rule. To begin with, Third World countries have unimaginably tiny amounts to spend on public health. (The poorer African countries spend eighteen dollars per person a year on health. We spend five thousand dollars.) And then the market value of a vaccine may be a fraction of its *social* value. If you're vaccinated, it not only makes you safer; it makes me and my children safer, too.[2]

What Gates and others are nurturing is a new kind of non-profit that strategically repositions itself and directs its energies toward long-term market solutions. By this logic, an NGO, or nongovernmental organization, becomes an MDO, or "market-directed organization." Hoffman has positioned Sanaria as just such an enterprise. Even though it is a for-profit biotech company, it has been capitalized by philanthropic dollars. Its future investment is dependent not only on effective clinical trial results but on the kind of profitable market, made up of military personnel, tourists, and business travelers, that Hoffman estimates to be worth billions of dollars.

Some nonprofits are taking a less extreme path. To supplement their fundraising, they are leveraging their assets and creating profitable revenue streams through the sale of goods and services, engaging in cause-related marketing partnerships. A market-directed organization goes beyond that trend, taking steps that make it more conducive for markets to act responsively toward it. They are developing metrics to assess return on investment and instill accountability, for example, and ensuring greater transparency about impact. They are also pursuing financing strategies that include the use of debt and equity and paying attention to long-term capacity building to create built-to-last organizations.

For even the most well-intentioned NGOs, none of this comes naturally. But a variety of strategies has already emerged from the new breed of nonprofits remaking themselves as MDOs. In addition to the efforts described above, these include:

- Fee-for-service: College Summit helps talented low-income students navigate the college access process. Valuable to school districts that compete for the best ratios of college-going graduates, this national nonprofit organization created a new model for nonprofits: Philanthropy covers its national overhead costs, but fees paid by schools cover local costs as the organization expands into new communities to provide services. This market mechanism enables College Summit to grow without spiraling dependence on charitable gifts.
- Creation of community wealth: More organizations are leveraging their assets to generate earned revenues. In Baltimore, the Caroline Center, which provides job training for women, runs a furniture-reupholstering business. In Rockville, Maryland, the Jewish Social Services Agency runs a private-duty home-health-care service. Each generates new revenues to support its nonprofit mission.
- Public/private partnerships: The Reinvestment Fund, The Food Trust, and the Greater Philadelphia Urban Affairs Coalition created the Pennsylvania Fresh Food Financing Initiative to increase the number of supermarkets offering fresh and nutritious foods to underserved communities whose needs would not be met through conventional financing institutions A $30 million appropriation from the state legislature was combined with private resources from The Reinvestment Fund to establish a grant and loan program that

has provided funding for eighty-three supermarkets in thirty-four Pennsylvania counties. It is an excellent example of political, nonprofit, and market forces combining to serve a public need.

- Investment in entrepreneurs: The Acumen Fund is a nonprofit that invests philanthropic dollars in business enterprises that deliver affordable and critical goods and services in poor countries. Investments include Kashf, a commercially viable financial services company making small loans to women in Pakistan, and the A to Z Textile Mill in Tanzania, a privately owned company employing 2,000 people that produces insecticide-treated bed nets to prevent malaria. An initial investment of $375,000 by Acumen enabled A to Z to scale up production and lower its costs per net. It is now the largest bed-net manufacturer in Africa.

Each of these efforts is very different from the others, and yet there is a common thread. What they all share is the attempt to redress the shortcomings of the philanthropic marketplace—the fact that philanthropy typically is not responsive to performance in the same way that capital markets are. Other mechanisms must be developed if nonprofits are to be responsive, and so, as Gates said, we must "stretch the reach of market forces" to address social needs.

Not every nonprofit needs to concern itself with transforming into an MDO. Many small, local efforts to expand a cultural institution, to promote better environmental practices, or to support neighbors in need could well thrive on charitable support alone. But when it comes to solving complex national and international problems, addressing root causes and not just symptoms, then some alignment with market forces designed to attract more capital is likely to be a necessity.

VENTURE CAPITAL FOR VACCINATIONS

To create the malaria vaccine that has long eluded tropical disease specialists, it is not enough to be an expert in infectious disease, parisitology, molecular biology, and genetic engineering, or even to have assembled the best talent in each of those fields. Nor is it enough to have access to the best laboratory and research equipment. Medical knowledge and scientific achievement represent just one aspect of what is needed.

At an initial cost of $400 million to $500 million to bring a vaccine from laboratory discovery through clinical trials and production, vaccine development is big business. As such, it requires—in addition to scientific knowledge—expertise in corporate finance, venture capital, IPOs, engineering and manufacturing, and supply-chain management. The physicians and researchers who care about vaccines for neglected diseases and drive their development almost never have training in those disciplines. Steve Hoffman is no exception. That is what brought him, before his Gates funding came through, to the Madison Avenue office of legendary investor and philanthropist Ray Chambers on a sunny spring morning in New York City. That, and the fact that by December 2009 he was expecting to run out of money.

Ray Chambers had made his fortune at Wesray Capital, the New York City firm where he had masterminded the leveraged buy-outs of Gibson Greeting Card and Avis Rent-A-Car in the early 1980s. He had then turned his attention to philanthropy, both in his hometown of Newark and around the world. He had a penchant for investing in for-profit activities that had a social impact, precisely because they were more sustainable as for-profits.

It was a big week for Steve Hoffman. That night he expected to see his colleague, malaria researcher Pedro Alonso, who was in the United States for a visit from his post in Barcelona. Alonso was then the lead investigator for the clinical trials on the RTS,S vaccine. The results of the Mozambique trials were considered promising, although almost everyone in the field found them to be preliminary, at best. Nevertheless, it was the hot prospect, generating press and buzz and dollars.

Hoffman had known Alonso for a long time, and it was clear he respected Alonso's ability to command so much attention and support. "He's a great medical and scientific entrepreneur," Hoffman told me. "But I think what Pedro really wants is to test our vaccine. He knows RTS,S is a first-generation vaccine that is only partially effective. So when people ask what I think about their trials and the vaccine clinics they are setting up with the $105 million grant from Gates, I say, it's great, because those clinics will be there for others to use later."

Ray Chambers, immaculate, as always, in a black suit, sat at attention, put his hands together, and listened carefully as Hoffman told him about his background in the navy and explained how he'd moved to Indonesia to work at the navy's research lab in Jakarta. "After having the experience that many tropical disease doctors have, of having many children die in my arms, I decided to focus on development of a vaccine," Hoffman told him.

Hoffman described his vaccine work and spoke of being recruited by Craig Venter to work at Celera, which he thought would be of interest to Chambers. The fact that Steve had been part of the ambitious effort to map the human genome caught Ray's attention. The discussion shifted from science to business as Hoffman explained the advice he had received to make Sanaria a for-profit enterprise. He spelled out his strategy to develop a first-world vaccine that would be available at a 95 percent discount to the developing world: "There are between 10 and 15 million travelers between the U.S., Europe, and Asia. Even with inept marketing and just a 20 percent penetration rate, a vaccine could bring in three-quarters of a billion dollars," he said.

The problem was that the foundations didn't want to participate in the development of a vaccine for the developed world. They wanted their dollars to be targeted exclusively to a vaccine that would reach the infants in Africa.

"In the entire history of vaccines there has never been a vaccine used in the developing world that was not a first-world vaccine first," Hoffman explained. "You can't go to

Africa and meet with prime ministers and their health advisers and say, 'We have this vaccine that is good enough for you but we've never put it into our own children and we're not going to.' That just doesn't fly."

Ray Chambers offered lots of advice supporting Steve's instinct not to give up equity, and therefore control, to a larger biotech VC firm at this early stage. He agreed that the for-profit model was the best way to bring production to scale. "This is a conversation I have with myself every day," Chambers said. "Can I do more for the world if I keep making more money?"

It could take $500 million to get the vaccine through licensing, but Hoffman believed that much of that amount could come through private capital, once "proof of principal" was established. He needed $2 million to $6 million in the short term, but even $1 million would be valuable, as he was spending $400,000 a month and bringing in only $200,000. The toxicology studies alone would cost $800,000. The expense resulted largely from the massive amounts of paperwork required to meet FDA documentation requirements.

As the meeting went on, it became clear that the questions confronting Steve, the unknowns, were not about molecular biology or chemical reactions, but about financing and logistics. He was schooled in the former. The latter would have required a Stanford Business School MBA to master. How to finance a $500 million enterprise? What types of venture capital to take? How much equity to give up? What financing mechanisms were right for the various stages of production and manufacturing? When he did experiments in the lab, he could repeat them if they went bad. With financing decisions, he risked getting locked into an irreversible course, especially when it came to giving up control.

Ray asked Steve questions about his production plans and the timing of the capital needed. Hoffman said he had enough money for a few more months, but he was determined not to allow short-term needs to compromise his ability to make decisions in the long term, or to let foundation program officers with two years of biotech "experience" decide how his vaccine should be developed.

Ray passed along advice he'd received from a Buddhist, though it was a Latin proverb—*festine lente*, "hasten slowly"—and then brought in his chief investment officer, Carla Skadinsky. She had been with him for five years but before that was at Goldman Sachs for twenty years and was at Rockefeller University in between. She suggested several names of potential investors and promised to look at any materials Hoffman would send along.

Afterward, Hoffman reflected, "Everyone in this field, whether the pharmaceuticals or the foundations, think they know best and want to do it their way, and I guess I do, too, I think I know what's best, and I've come too far to walk away from that now. But it's never easy. There's always something, you just have to keep pushing forward."

On February 14, 2008, United Nations Secretary General Ban Ki-moon appointed Ray Chambers as the UN's first special envoy for malaria. For the previous eighteen months, Ray had been the cochairman of an organization he founded called Malaria No More, which worked to raise funds and create awareness to combat the disease.

I had met Ray in 1996, when he was involved in organizing what was known as the Presidents' Summit with retired general Colin Powell. That was in the years before Powell became secretary of state for President George W. Bush, and the summit convened for three days in Philadelphia in the spring of 1996. Along with all of the living presidents, representatives of businesses and organizations announced commitments to get necessary resources to America's poorest kids. It led to the creation of "America's Promise," a foundation supporting "Five Promises" to youth to help them succeed.

Ray had been devoting himself to mobilizing commitments on behalf of kids for years, ever since leaving the leveraged buy-out industry that he had helped to create. As his passions evolved, he had already begun to support mentoring—the focus of one of the Five Promises—seeing it as the kind of silver bullet that could change children's lives. In that spirit, he had taken up the idea of a summit of all of the living presidents and run with it. His primary purpose was to call attention to the resources children needed.

As his interests turned from revitalizing Newark to a more national and global outlook, he had asked if I would introduce him to Jeffrey Sachs, the Columbia University professor, author, and anti-poverty advocate. That led to Ray's embrace of the Millennium Development goals and the creation of yet another organization, called Millennium

Promise. The goals are a set of eight globally endorsed objectives—such as providing universal primary education and halting the spread of HIV/AIDS—intended to address the multiple causes of extreme poverty, and Millennium Promise mobilizes both public and private partners to achieve them. Ray traveled to Africa and saw children dying of malaria, and his commitment to doing something about the disease became irreversible.

Ray is particularly good at bootstrapping his way forward, identifying someone whose support will bring further credibility, and then parlaying a small achievement into larger ones.

This role of UN envoy is his most public to date. Before that appointment, Ray had always worked behind the scenes. But unlike most philanthropists, he understands that big change and big results don't come cheap. He is willing to pay to have the best people in any field be a part of what he is doing. "Belts and suspenders, Billy, belts and suspenders," he used to rasp to me over the phone as a way of saying "spare no expense." Ray has not only devoted a substantial portion of his personal fortune to philanthropy but personally works far harder than he has to, spending long days learning from experts and pursuing and cajoling others with a proven ability to accomplish things to join in his efforts.

Ray's good intentions have at times been undermined by the seduction of a too simple solution—such as increased distribution of bed nets, which, while vital to saving lives, can't by itself end malaria. But as a man of almost unlimited imagination, he is attracted to the challenge of solving big, complex problems that intimidate others, and he brings a vision and expansive sense of what is possible to those tasks that others often lack.

"AT WHAT POINT WILL WE TREAT IMMUNIZATION THE WAY WE TREAT UTILITIES?"

In October 2007, the *New York Times* published a story in its science section describing how scientists fighting malaria were split over the best strategy for distributing insecticide-treated bed nets to help prevent the disease. The article unintentionally said as much about the state of the nonprofit sector today as it did about science, highlighting an ongoing debate about how the tools and assets of government, philanthropy, and the marketplace might complement each other to solve social problems. The *Times* reported:

> Recently, Dr. Arata Kochi, the blunt new director of the World Health Organization's malaria program, declared that . . . the only way to get the nets to poor people . . . is to hand out millions free.
>
> In doing so, Dr. Kochi turned his back on an alternative long favored by the Clinton and Bush administrations— distribution by so-called social marketing, in which mosquito nets are sold through local shops at low, subsidized prices . . . with donors underwriting the losses and paying consultants to come up with brand names and advertise the nets.[3]

I raised the issue the day after the article appeared during dinner in Barcelona with Pedro Alonso, one of the world's largest and most grateful beneficiaries of American philanthropy. Tropical medicine's equivalent of a rags-to-riches story, Alonso was to join Bill Gates in Seattle the following week to announce another milestone in saving the lives of the world's poorest children.

Like Hoffman, Alonso is an alumnus of the London School of Hygiene and Tropical Medicine. His primary partner in fighting malaria has been his wife, Clara, whom he met in school. They have three children. Alonso views malaria as both a cause and consequence of poverty in the most undeveloped regions of the world. Based at the University of Barcelona, where he has taught since 1992, he also leads Barcelona's Center for International Health Research, where his efforts are part of every discussion about the need for a vaccine.

Under the best of circumstances, vaccine development takes extraordinary amounts of money, time, patience, commitment, and luck. The few who have succeeded are like Olympic gold medal athletes who graduated summa cum laude and also happened to win the Powerball lottery. Alonso is optimistic but also realistic. Writing in 2006, he explained:

> One of the greatest problems with malaria, which accounts for the extreme difficulty of developing a vaccine, is that we do not yet understand how individuals develop immu-

nity against the disease. Since . . . the end of the 19th century, it has been known that adults who survive malaria
infection acquire a highly effective immunity, but the
mechanisms involved and how they operate remain unknown, as is the role of indirect protective measures. . . .
Thus, the only way to assess whether a vaccine is effective
is by conducting clinical trials in natural conditions.[4]

In 1996 Alonso built a sophisticated medical facility,
the Manhica Health Research Center, in one of the most
malaria-infested areas of Mozambique for this purpose.
Mozambicans have an average life expectancy of just forty-
eight, and there are only 800 doctors for a population of
18.9 million.[5]

Alonso is credited with putting Barcelona on the map
as an international center for global health research. I expected his office to be something more than the nondescript fourth-floor walkup pointed out by a bored shop
employee in the lobby of a building near the fashionable
shopping district of Barcelona. Nevertheless, I went up for
our 4:30 appointment, and an assistant told me Alonso was
in a meeting. I waited, but 4:30 came and went, as did 5:00
and 5:15. Finally he rushed into the room through a nearby
door, shook my hand, said, "Sorry, I'm in a conference,"
and disappeared again.

The office looked like that of many mission-driven nonprofits, with walls sporting maps of Africa and photos of
children at medical clinics in small villages. It was quiet,

with most of the staff staring at computer screens. At one point Alonso came out again to take a call on his cellphone, saying only, "Sí, sí, sí, sí, Gates, sí, sí, sí, sí," before returning behind closed doors.

When he finally emerged an hour past our scheduled time, I guessed, accurately, that he was late for his next appointment. He asked how long I was in Barcelona and whether we might have a drink or dinner. A quick call to his wife confirmed that she could join us. He returned, said, "Well then, it's all arranged. See you tonight." With that, Alonso turned and the encounter was over.

Despite the delays, I found Alonso impossible not to like. Intense but relaxed, he resembles a more sober version of Dan Ackroyd, with a young Ackroyd's dark hair and the middle-aged Ackroyd's girth. He speaks excellent, authoritative English in a deep resonant voice that could make him an anchorman on an English-language news network.

An exchange program with Boston University had brought Alonso and his wife to the United States and to Boston City Hospital, where they had found mentors who steered them toward international health. Clara specializes in malaria control in pregnant women. When asked how they both ended up working on malaria, Alonso told me: "The real question is how can you not work on malaria? If you are interested in African health and poverty, everything depends on addressing malaria."

Their decades of travel to Africa, the research facilities they built, and the people they trained in Tanzania, Mozambique, and the Gambia had by 2003 created an infrastructure in African villages that the Gates Foundation found appealing and rare. Speaking of Bill and Melinda Gates, Alonso said, "They got it. Gates himself got it. They got that vaccines in Africa are not just about science and developing the product. The question is always what do you have on the ground? How do you test it? What about infrastructure?"

The commitment of the Gates Foundation to vaccine development for malaria has included hundreds of millions of dollars to support numerous competing vaccine development efforts. More than $107 million had been initially earmarked for Alonso's work. Hundreds of millions more dollars have gone into refining RTS,S. As lead investigator of the clinical trials for RTS,S, Pedro Alonso quickly became the public face of malaria vaccine development.

The formidable hurdles to vaccine development usually cause the hurdlers to become fierce competitors, but Alonso seems to take a longer and more generous view. "RTS,S will likely be the first malaria vaccine to get approved," he told me, "but it won't be the final vaccine. They will continue to evolve and develop, and RTS,S will be a contributor to it. Actually the wild idea of Steve Hoffman may make the best vaccine, but he still has challenges to overcome."

Unlike Steve Hoffman, who sees tourism and the military creating a market that would help pay for his vaccine, or Jay Keasling, who is allowing synthetic fuel development interests to subsidize his development of bioengineered artemisinin, Alonso does not talk much about market mechanisms.

Just the opposite, actually. In 2008 Alonso told a reporter for GlobalPost, "It's pretty simple . . . If the head [of a pharmaceutical company] gets up this morning and announces that they're going to invest $1 billion [in a malaria vaccine], by lunchtime . . . the stock would have plummeted and the guy would be out. Who is going to buy a malaria vaccine?"[6]

In fact, Alonso is unabashed in his rejection of the social-marketing approach. "This is something we simply must do," he told me, referring to the project of making malaria nets and vaccines available to all who need them. "At what point will we treat immunization the way we treat utilities? The government sees the public health benefit of getting clean water to you. How will it be different for a malaria vaccine? We must see that it will take fifty to one hundred years to achieve this and be willing to make that commitment," he said, calling to mind the cathedral builders in this city that is home to the world's most famous unfinished cathedral, Gaudi's Sagrada Familia, 150 years under construction and still lacking a finished roof.

But what if political will is lacking? What if it is just not in the cards politically? Wouldn't you want to find market mechanisms that could be put to use to achieve at least some of your objectives? I asked Alonso these questions. "That would be okay," Alonso conceded reluctantly. But he believes the market would not be able to accomplish the task as resolutely or as completely.

The debate is emblematic of the continuing evolution of the nonprofit sector. The same issues underpin so many of the developments taking place today. If it feels like the non-profit sector is in the midst of turbulent transformation, especially in the United States, it is because the demands placed upon it have moved it ever closer to the fault line that has defined American politics since at least World War II, a fault line dividing those who believe it is government's responsibility to improve the lives of individuals, and to help to care for those who cannot care for themselves, and those who believe such tasks should be left to the individuals themselves, their families, and their communities, aided by whatever incentives the market sees to join that effort. Few issues, outside of religion, evoke such deeply held beliefs.

Whether it is Gates's creative capitalism philosophy or Alonso's that prevails, the debate is likely irrelevant to the African mother of a four-year-old girl convulsing from the fevers and chills of severe malaria. She is less political than pragmatic, wanting only whatever will work to save her child. Saving her, and millions like her, requires society to continue to struggle to find a balance between political solutions and market solutions, and not to exclude either completely. For that essential underlying reason, Gates, despite his belief in creative capitalism, has poured enormous philanthropic resources into Alonso's work. Gates and Alonso are not letting the politics of the market divide them from the task at hand: eradication.

CHAPTER 10

MOMENT OF TRUTH

Army medical researchers reported today a major breakthrough in the fight against malaria, next to the Viet Cong[,] the most savage enemy American troops face in Vietnam.

A spokesman for the Surgeon General's office said diaminodiphenylsulfone (DDS), a drug long used in treating leprosy, was found in Vietnam field tests to cut in half the number of men who are stricken by malaria. . . .

. . . The Army considers the malaria problem urgent enough that necessary final approval from the Food and Drug Administration was arranged in a telephone call.

—Associated Press, June 23, 1966

IMAGINE CHECKING INTO A HOTEL for a week-long stay on a sunny spring morning. After the receptionist at the front desk greets you, checks you in, and offers a key to the mini-bar, you go to your room, unpack your suitcase, and open the curtains to check the view. You change into comfortable, loose-fitting clothes. But instead of heading off to a conference or to the fitness center, you walk down the hall to a room and knock on a door, which is opened by a military official. He directs you in and toward a chair. Once you are seated, he awkwardly straps a cylindrical white box onto your arm. The box is swarming with mosquitoes. They not only take bite after bite of the tender skin on your forearm, but have been bred to inject into your bloodstream, along with their saliva, malaria parasites of the species *Plasmodium falciparum*, the same parasites that kill 1 million children and sicken another 300 million people worldwide every year.

This is what it means to be "challenged" in a clinical trial for the volunteers who meet the strict criteria for participation, and were previously inoculated with an experimental malaria vaccine candidate. The hotel is filled with such volunteers, as well as doctors and nurses who take blood samples from them once a day and perform exams. Anyone presenting with evidence of malaria, measured in numbers of parasites per microliter of blood—can be treated at once with powerful antimalarial drugs, even before symptoms set in, so that the parasites are permanently cleared from their bloodstream. It is standard procedure and perfectly safe. Since no vaccine to prevent malaria has ever been fully effective, though, it is a procedure that has been necessary to follow time and time again to return challenged volunteers to good health.

On Tuesday, May 26, 2009, Steve Hoffman began Phase I of just such a clinical trial in Maryland on the vaccine he had dedicated much of his life to developing. For more than a year, Hoffman had been telling anyone who would listen that "trials will start in the next month or so." But like the horizon, the date always seemed to recede as he moved closer to it. There was always some complication or delay, most of them probably fully anticipated by Hoffman in the first place. But he understood that dangling the imminence of clinical trials, like dangling a carrot in front of a horse, would keep things moving. The nature of science and experimentation is, in part, making it up as you go, and like a successful TV talk-show host, Hoffman knew that among his most useful injunctions were "stay tuned" and "coming up next. . . . "

The start of the trial was as much personal triumph as professional milestone. It was a day that the vast majority of experts in the malaria field thought would never come. Some had said so publicly, while others had snickered privately. Many simply ignored Hoffman, focusing their energies instead on more conventional approaches that were farther along than his. In the press release that Sanaria and the PATH Malaria Vaccine Initiative put out on April 23, 2009, Myron M. Levine, director of the University of Maryland School of Medicine's Center for Vaccine Development—one of the locations for the trials—noted that Hoffman's vaccine was based on studies from the 1970s "that were never translated into vaccine development effort because the task was considered to be impossible."[1]

But Hoffman had persevered, undeterred, if not unfazed. He had reached this day thanks to a combination of paradoxical conditions: personal financial sacrifice that cost him more than a few nights' sleep, until earning generous philanthropic support that restored his peace of mind; a devoted following of true believers committed to disproving the skeptics, who vastly outnumbered them; and a brilliant wife and partner who kept a low profile but whose own discoveries had made key stages of his work possible.

During that last week of May and the beginning of June, the team initiated inoculations of 104 people via four injections at two sites. One was Levine's Center for Vaccine Development; the other was the Clinical Trials Center at the National Naval Medical Center in Bethesda, Maryland. In September the two sites began the "hotel challenge." The verdict on the trial would take six months to compile. If research determined that the vaccine was safe, Hoffman could proceed with further phases of testing and trials to assess effectiveness. There would be a need for additional large sums of money, but no one is better than Hoffman when it comes to parlaying small victories into a larger narrative that inspires further support.

By coincidence, or perhaps not, on the same day that Steve Hoffman's vaccine began clinical trials in Maryland, his principal competitor, GlaxoSmithKline, was commencing more advanced trials 8,700 miles away in a small town called Bagamayo, a once important trading port and now a center for sailboat construction on the east central coast of Tanzania. Five infants between the ages of five and seventeen months were being inoculated with GSK's RTS,S vaccine.

RTS,S was advanced enough to be launching Phase III of its clinical trial, the largest ever of a malaria vaccine. The five infants were the first of 16,000 kids in seven African nations in a trial that would last seventeen months. Half would get RTS,S, and half a placebo. All would be given insecticide-treated bed nets. They would then be clinically monitored for two years.[2]

When Phase III is over, RTS,S will either be crowned as the great new hope for the African continent, and rushed into production at GSK's Rixensart, Belgium, manufacturing facility, or it will go back to the lab for refinement. Perhaps both. RTS,S is widely expected to be the vaccine that is first to market, but not the final word. It appears to only provide protection to about half of those vaccinated. Critics, including Hoffman, argue that it doesn't prevent infection so much as delay its onset. Even so, that may be enough to allow small children to survive until their own natural immunities kick in to make them less vulnerable. They would still contract malaria, but the hope is that it would be less debilitating.

RTS,S is, to date, the only malaria vaccine candidate, out of more than seventy that have fallen short, to get as far as Phase III trials. As such, the launch in Bagamayo, a town of only 30,000, got extensive media attention, internationally as well as in Hoffman's own backyard, and especially in the science and medical trade press. American, European, and African newspapers and websites were filled with reports of this unprecedented undertaking. The coverage threatened to swamp Steve Hoffman's efforts.

RTS,S, the kind of high-tech, subunit recombinant DNA vaccine in vogue today, has an "adjuvant" designed to boost its effectiveness (in this case a surface antigen of the hepatitis B virus). Hoffman's vaccine is a throwback to the earliest days of vaccine development: a vaccine composed wholly of the very parasite (weakened but still strong enough to trigger the immune system) that one intends to destroy, similar in premise to the vaccines for the smallpox and polio viruses.

The automated manufacture of RTS,S is capable of being scaled up in mass quantities at GSK's sophisticated production facilities. Hoffman's vaccine requires the labor-intensive costs of lab technicians doing precision piecework, the equivalent of a shoe manufacturer like Timberland making expensive hand-sewn boots in the Dominican Republic when many of its competitors use technologically advanced factories in China.

If either vaccine proves successful—and leads to the eradication of malaria—it would be only the second time in human history that a disease has been wiped out. The first was the elimination of smallpox in 1979. Even polio, for which a vaccine was discovered in 1952, was diagnosed 1,300 times in 2007 and poses the threat of resurgence.[3]

Getting to the RTS,S trials had taken GSK decades and cost tens of millions of dollars, with the team behind the project enduring dozens of conferences, writing dozens more peer-reviewed papers, and going through all the requisite disappointing experiments, regulatory approvals, and more. Getting through the trials may be even harder.

Just a few days after the RTS,S inoculations began, the *Toronto Globe and Mail* described the test as a task "that pits hundreds of scientists, doctors and field workers against floods, famine and corruption that corrode the very core of health infrastructure and breeds deep public distrust."[4]

In Kilifi, Kenya, the skepticism of tribal chiefs and mothers was so deep that it took more than six months to recruit just four hundred volunteers. Daunting logistical issues compounded the challenge: The *Globe and Mail* reported that, "As the long rains set in, trucks wallowed and got stuck in thigh-deep mud, everything from samples in need of refrigeration to sick children had to be moved across dozens of kilometres by motorbike or on foot. Field workers in rubber boots struggled to access homesteads, some of them mere islands in a sea of mud."[5]

This is what the makers of RTS,S are dealing with today. It's what Hoffman hopes he gets far enough to deal with tomorrow.

ONE STEP FORWARD
AND ONE STEP BACK

The milestone of clinical vaccine trials finally achieved by Hoffman as well as RTS,S is especially timely in light of new and alarming reports of growing resistance to the one drug that until now has been the most effective and reliable method for treating malaria, artemisinin. It is a development that could also drastically undermine the impact Jay

Keasling seeks to have by using synthetic biology to inexpensively increase supply.

The emerging resistance to artemisinin is partly due to the complexity of the parasite and the unstoppable nature of its ability to evolve. But it is also the direct result of greed, for which there is no drug or vaccine.

In early 2009 the World Health Organization (WHO) announced the discovery of malaria parasites along the Thai-Cambodia border that were proving resistant to commonly used derivatives of artemisinin called artesunate and arthemeter.

Prior to this there had been a hugely optimistic surge of interest, research, and spending on combination therapies using artemisinin. Chinese companies developed numerous forms of artemisinin and combined it with another anti-malarial drug called lumefatrine so that the parasites would be less likely to develop resistance. China's own malaria caseload went from 2 million in 1980 to 90,000 in 1990. Colonel Peter Weina, who, as chief of the pharmacology department at Walter Reed Army Institute of Research from 2002 to 2009 had led the U.S. Army's effort to develop a drug against the most severe form of malaria, went so far as to admit that, "there was an assumption and there was a hope and there was a prayer that resistance to artemisinin would never happen."[6]

History teaches that we should have known better. Resistance to malaria medicines has always been a question of when, not if. In previous decades malaria had become resistant to once potent drugs like chloroquine in the 1960s and mefloquine by the 1990s. A category of drugs known as SPs (for sulfadoxine pyrimethamine) was introduced in 1977 with a 100 percent success rate. Though it remains widely effective in Africa, within five years SPs were effective in only 10 percent of patients in some areas of Southeast Asia.

Wishful thinking held that artemisinin might be the exception. But the new discoveries are showing that artemisinin, which once cleared parasites from the blood in two days, is taking three or even five days in a large number of those infected, the first sign of growing resistance.

The Chinese sold the rights to the drug to Novartis in 1994. Novartis sold it at $44 a course to Western travelers but agreed to sell it to the World Health Organization at $2 a course. Even at that price it was ten to twenty times more expensive than previous drugs, such as chloroquine, and USAID and CDC officials considered it too expensive for Africa. Then, an underground market developed. Underground markets, by definition, are unregulated. As a result bad things happen.

One of those bad things was counterfeit drugs. Paul Newton, head of the Wellcome Trust Southeast Asian unit in Laos, concluded that tablets with low doses of artesunate—insufficient to kill the parasites, but enough to lead to resistance—were the main source of the growing problem. Pills with only small quantities of the active ingredient, instead of the full amount, are dangerous because they wipe out only the weakest parasites, enabling the hardiest to survive and multiply. When a malaria victim infected with resistant parasites is then bitten by another mosquito, and that mosquito bites another human, the resistant parasites have a chance to thrive and multiply some more. And on and on it goes, like a breeding plan for more and more drug-resistant parasites. After a while, the genuine drugs will no longer work.

The pirated drugs are often sold in packaging indistinguishable from the real medicine and contain small amounts of active antimalarial substances in order to pass quality-control tests. Studies have found that between one-third and one-half of artesunate tablets across Southeast Asia were counterfeits. The criminal trade in these cheap imitations was made attractive by the relatively high cost of the real drug—about $2 for a course of treatment, or up to $10 in the private sector. A course of chloroquine, by contrast, costs only 10 cents.

Another contributing factor was the widespread availability in the region of high-quality artesunate tablets from China and Vietnam. The artesunate in these tablets was present as a single ingredient, rather than in a combination of ingredients, which also gave the parasite a better chance of developing resistance.

In addition, because of the relatively high cost of the drugs, poor people were often not completing their courses— they stopped taking the tablets when they felt better, giving any resistant parasites a chance to survive and proliferate.

In 2008, the International Criminal Police Organization (Interpol) arrested twenty-seven people in raids across Asia and seized 16 million counterfeit pills worth at least $7 million. The global market for fake pharmaceuticals is estimated to be $75 billion. As Paul Newton of Wellcome Trust has pointed out, the crime really involves more than just making cheap drugs: "If you make a medicine that contains no active ingredient for a disease you know can be fatal," he said, "at best that is manslaughter and at worst it is murder."[7]

The crime is even greater when one considers all the consequences. On one level, selling the pirated drug could easily cause one person to die—the person who took that drug and as a result did not get better. But on another level, selling the pirated drug contributes potentially to the death of millions, since the imitations are causing the real artemisinin drugs to become ineffective. "Twice in the past, South East Asia has made a gift, unwittingly, of drug resistant parasites to the rest of the world, in particular to Africa," according to Nick Day, director of the Mahidol-Oxford Tropical Medicine Research Unit. "If the same thing happens again . . . that will have devastating consequences for malaria control."[8]

Virtually all malaria researchers would concur with Day on that point. "If we lose the artemisinins at this stage, just now when we dare to mention the word 'eradication' again, it would be a disaster for malaria control," said Dr. Arjen Dondorp, lead author of a study about artemisinin resistance in Cambodia. "It would cause millions of deaths, without exaggeration."[9]

There is yet another level of danger arising from these circumstances, with the consequences going far beyond even losing the benefits of a once-effective medicine, as serious as that would be. The greater problem is that the parasite could come back stronger than ever. Partial victory against malaria could ultimately be worse than total failure.

History offers one such example that traumatized the malaria community for nearly half a century. In 1955, the World Health Organization believed that malaria could be eliminated within ten years, thanks to the introduction of chloroquine in the 1940s and the discovery of DDT, the most effective insecticide in history, in 1939. Today we are more aware of the dangers of DDT than its benefits, but in 1948 the chemist who created it, Paul Müller, won the Nobel Prize in medicine for doing so.

WHO spent more than a billion dollars in its campaign against malaria, and the disease was wiped out in much of the Caribbean and South Pacific. Malaria cases in Sri Lanka dropped from 2.8 million in 1946 to a total of 17 in 1963. Likewise, in India there were 75 million cases in 1951 and only 50,000 in 1961.

But in the deep tropics, the disease remained untouched, and the program largely bypassed sub-Saharan Africa. Overuse of DDT by farmers (not by malaria fighters) led Rachel Carson, who is widely considered the founder of the environmental movement, to document its abuse in her book *Silent Spring*, and as a result of her revelations, much of the world outlawed it for agricultural use. DDT became politically incorrect and difficult to procure. The political will for financing malaria eradication eventually began to fade, and resistance to chloroquine created epidemics of malaria even more difficult to treat. The disease came roaring back in India and Sri Lanka. Today the toll from malaria is almost twice what it was a generation ago.

So when a leading drug such as artemisinin, once considered infallible, begins to falter, attention shifts to other approaches. One result has been to look to vaccines as a promising alternative to drugs. Hence the intense competition among leaders in the race to a vaccine, like Steve Hoffman and GSK. But there is also competition from advocates of other strategies for defeating malaria.

Just as different branches of the armed services employ different weapons and tactics to achieve the same goal, the malaria community has always been divided into different branches to wage its war, and each branch vigorously champions its own particular capabilities and worldview. Vaccine developers believe that prevention would be the most effective and economical method of combating malaria. Without a vaccine, they argue, we will be trapped in a perpetual cycle, spending massive amounts of money and energy on treatments that the parasite will eventually be able to resist.

Those who develop drugs to treat the disease think a vaccine would be great but quickly point out that there has never been one for malaria, or for any other parasitic disease. So the old debate begins again: Do we encourage palliative treatment today that falls short of eradication, or support a total cure that is years (and millions of deaths) away from universal application, if indeed it works?

And there are those in the malaria community who believe that resistance to both drugs and vaccines is just a matter of time and that we should focus on the only thing that has ever really worked: "vector control," that is, using bed nets and insecticide sprays as the weapons of choice. Thus we would try to get rid of the malaria parasite by getting rid of the mosquitoes that house, transport, and deliver it. But even this option is not exempt from evolution's toll. At the Fifth Multilateral Initiative on Malaria, held in Kenya in October 2009, it was reported that mosquitoes were adapting to bed nets by changing their feeding habits. Instead of trying to feed at night, when people were safely beneath their nets, the mosquitoes were getting their meals earlier in the evening when people were still out in the open. At the same time, there is evidence of the mosquito developing resistance to the pyrethrum-based insecticides used to treat the nets.

Practitioners within any one branch tend to be consumed by developments in their own field and not especially knowledgeable about the others. They can be uninterested at best, and dismissive at worst. Steve Hoffman and Jay Keasling devote to each other, at most, a glimmer of recognition—and less than a glimmer of enthusiasm. Or maybe it is simply a matter of staying focused on what one does best. As Hoffman told me of Keasling's work: "I admire what he's done, but I leave his job to him and he leaves mine to me." In each branch of the malaria war, there are many who believe their own approach embodies the best mix of compassion, realism, and effectiveness, and therefore occupies the moral high ground.

New discoveries, technological advances, and policy shifts regularly reset the playing field, temporarily boosting one approach over another and causing tidal shifts in the flow of resources, money, and manpower from one branch to another. Over time, however, it has become apparent that the malaria parasite is too complex and evasive to succumb to the attacks of one branch alone. All are necessary.

"All" gets very expensive, very quickly. Even more, it creates a challenge: the need for scientists to coordinate with each other and work together. The various branches of our military forces have rarely been successful any other way. The various branches of the malaria community have rarely even tried.

CAREFULLY CHOREOGRAPHED CLINICAL TRIALS

I first heard about Judith Epstein, a U.S. Navy commander, from Steve Hoffman when he told me about the editorial she'd written in 2007 for the medical journal *The Lancet*, "What Will a Partly Protective Malaria Vaccine Mean to Mothers in Africa?," which raised some of the same questions that Hoffman has asked about the RTS,S vaccine. The kind of humanistic themes she brought out are rare in scientific journal articles written by military personnel. Epstein argued for a vaccine that went beyond the partial effectiveness of RTS,S, even though the military—at least the army—was deeply invested in it.[10]

Epstein works at the Naval Medical Research Center, which is located near the imposing Bethesda Naval Hospital. As principal investigator at the Navy's Center for Clinical Trials, Epstein was responsible for conducting and assessing the Phase I trials in 2009 of Steve Hoffman's vaccine. She will yield the first official evaluation of the safety and efficacy of the vaccine Hoffman has worked on for all these years, determining whether the more than $60 million that foundations and government agencies have already invested in Hoffman's effort will be followed by investments of even greater magnitude.[11] The trials are not "make-or-break" so much as potentially "break." It is only the first of numerous trials, but there is no Phase II without getting past Phase I.

I wanted to get to know Epstein, partly because a doctor who goes into tropical medicine sets himself or herself apart from 99 percent of the rest of the medical community. The difference goes beyond being smart, or altruistic, though the doctors who make this choice often are both. It has more to do with having made different choices about how to lead one's life, choices that isolate one from the mainstream in medicine and that result in a way of being that is, by nature, rather solitary. I wanted to know more about the quality of character that enables people like Hoffman and Epstein to make such choices, and I thought Judy Epstein might help me understand this. She did not disappoint.

I drove out to NMRC—known as "Namric" in military parlance—on a Monday morning and waited to clear a checkpoint manned by a squad of five or six military police.

I was directed to pull in behind a line of cars parked on the helipad, which was marked by an enormous white cross in front of the naval hospital. Behind it lay a naval base that looked more like a rural college campus, with old buildings, leafy paths, and the stately residences of admirals instead of the offices of a college president and dean. Epstein pulled up, rolled down her window, and told me to follow her. As we went through the checkpoint, where she returned the salute of a military police officer, and proceeded to the grounds of the Naval Medical Research Center, I noticed how carefully and thoroughly she obeyed speed limits, stop signs, and traffic directions. Though fifty-four and an officer for nearly ten years, Judy still acted as if she were new to the navy, too new to take casually the numerous rules and regulations that are second nature to military life.

Epstein would be easy to underestimate. She's small and slim, and she walks with a brittle gait because of an old back injury. When she got out of her van, I was surprised to find her in uniform: khakis, well-shined black patent leather dress shoes, and a black sweater with three gold stripes on each shoulder. As she got out of the car, she fitted a khaki garrison cap over a thick knot of black hair pinned up on the top of her head. Her demeanor was serious. Though she was friendly and open, I don't think I ever saw her smile.

We headed over to Building 141, the headquarters for the Center for Clinical Trials. An old whitewashed building from the 1940s or 1950s, it contained offices, conference rooms, and examining rooms. Epstein, who had an office in the back, introduced me to several other investigators, including a nurse and a doctor. The corridors were quiet. Some of the staff was in another building, busy copying 20,000 pages needed for an Institutional Review Board.

Judy came late to the malaria community—and even to science and medicine. Her father was a doctor, her mother a painter whose first husband was Will Barnet, an artist best known for his enigmatic portraits of women and girls. Notwithstanding childhood dreams of being a marine biologist or working on the *Hope* hospital ship, she left college after one year for ballet and joined the company of legendary choreographer Agnes de Mille. When the revival of *Oklahoma* opened at the Palace Theater on Broadway on December 13, 1979, she was in the cast.

It was the combination of her own back injury and de Mille's stroke that led Judy to leave dancing and become de Mille's caretaker. She grew more interested in health issues and considered becoming a physical therapist. At the age of twenty-eight she returned to Columbia University and "kind of locked myself away for three to four years." After graduating she took another year off because her father was battling colon cancer, and from there it was off to Harvard Medical School.

Interested in both parasitology and pediatrics, Epstein did a residency at Children's Hospital in Philadelphia and was then awarded a four-year research grant. In 1998, the last year of her fellowship, an adviser recommended that she join the navy, and she was given a billet in the dengue program. "Then I saw Steve Hoffman's papers and asked to meet him," she told me. "As soon as I met him I said, 'I have to work for him.'" She found a way to work *with* him instead, as principal investigator running the clinical trials that began in September 2009.

TRIALS AND TRIBULATIONS

The trial had two critical phases: the immunization, and the challenge. The threshold question for immunization was not the efficacy of the vaccine, but its safety. Would there be "breakthrough infections"? That is, might the vaccine be so strong that it not only triggers the immune system but goes so far as to make sick the person one is attempting to immunize?

Hoffman's wife, Kim Lee, oversaw the immunizations each day. At 6:00 A.M., she'd start the process of extracting the vaccine from the liquid nitrogen and then diluting it. It would be injected intradermally or subcutaneously, and then the volunteers would be observed for thirty minutes. There were four cohorts of about twenty people each.

Although every aspect of a clinical trial is planned out with great care, the human factor can play havoc with any effort to even approach choreography. Trials use volunteers, and though carefully screened, volunteers can still be unpredictable. For example, one man came down with fever, chills, and joint and muscle pains, which could have been a reaction to the vaccine that would stop the trial. But ten days of intensive diagnostic efforts showed that it was instead

Lyme disease. Low, medium, and high doses are administered, the high doses twice, and many things can go wrong. But two weeks later, Hoffman told me that "the headline is 'no breakthroughs'"—in other words, no infections.

But then Steve explained how, the day before they were to challenge the third group—"the day I'd been waiting for for seven years"—he

> got a call from Tom Richie at NIH, who said, "We've got a problem." All of the volunteers were fine but the Institutional Review Board had suspended the trial. . . . While no one had become sick or been harmed in any way, there were some issues around dosage that did not satisfy our standard of clinical practice, and so the trial was suspended. The IRB's first responsibility is to the volunteers and we all get that. It didn't matter that the vaccine worked or that no one got sick. The cost to us would potentially run to millions of dollars. This work is not for the faint of heart.

When I asked Hoffman how typical it was that a mistake like this would happen, he explained that, "of the 1,000 things that could result in a vaccine trial being suspended, this is one that I would never have imagined or seen coming."

In early 2010 Hoffman tried again, but by May he had more disappointing news to share: "There's a lot we don't know about how the mosquito delivers the parasites when it bites. In any case, what we learned is that our dose was way too low, more than tenfold too low. But that's what you do in a Phase I."

The results were illuminating, but a potential problem in terms of future funding: "We gave 80 volunteers one dose, 66 volunteers four doses, and 17 volunteers 6 doses," Hoffman told me:

> There were no breakthrough infections. It was safe and well tolerated. And that in itself is incredibly important. We then challenged them with bites from five infected mosquitoes three weeks after their last dose. But when we challenged group 1 there was no protection. In group 2 there were 2 of 16 protected. Group 3 could not be challenged. In a fourth group there was no protection. . . . One explanation was that we were not using enough parasites. Our senior advisers said we should be happy because we had proved what a Phase I sets out to prove, which is that the vaccine is safe.

Sanaria's main funder, the PATH Malaria Vaccine Initiative, he explained, "said we hadn't met their go–no go criteria and that, at least in the short run, there would be no more funds, despite the fact that more than $50 million had already been invested in this approach. . . . They wanted the clinical trials to show not only safety but effectiveness. So did I! But often that's not how science works. That's not what the first phase of clinical trials is for."

After all the years of preparing for that moment, all of the lab results, all of the papers, conferences, and collaborations, when it came to administering the vaccine Hoffman was left to make an educated guess. Trying intradermal and subcutaneous injections was a huge "maybe," and when the trial results finally came back, nature had answered "maybe not."

"This trial was what I'd been waiting for and working toward for almost a decade," Steve told me:

> I was devastated. It left me really morose—like stay-under-the-covers-and-read-trashy-novels-for-a-month morose. I thought, "I'm sixty-one, what the hell do I need this for? Maybe I should go practice family medicine in some small town in Idaho or Maine." It's what I used to do and I loved it, so I'd be okay with that. My dream of creating the vaccine that would eradicate malaria is over. I'm just another guy working in a lab somewhere trying to make a difference. And I'm okay with that too. That's reality.

After about two weeks, Hoffman finally pulled himself together and went to New York to see Ruth Nussenzweig. It was like journeying to Delphi to touch base with the oracle. "You know what she said?" he asked me. "She said, 'So you're no magician.' And she's right. I'm no magician. That's life. Who was I to think that on the first try I'd have the vaccine? That it would be easy for me and for me alone? In my head, I knew that you can't get that from Phase I. Still, I'd hoped."

He also went to see Tony Fauci at NIH, who said, "You must keep going!" And then he got a call from his son Seth at Cornell, who said, "Dad, it would be totally immoral and unethical to stop now, no matter how disappointed you are."

"Look. Bed nets aren't going to solve this problem," Steve said to me. People were dying, including Americans. He cited a Stanford student who had been traveling in Ghana, an American traveling in India, and a Seabee (a member of the Construction Battalion, or CB) from the U.S. Navy, all of whom had contracted malaria in recent months and died. "If I see you on the first or second day that you've been infected I can cure you for sure," Steve said. "The medicines work." But there are many who do not receive treatment until it is too late, if at all. At this point Hoffman pulled out a recent *Lancet* editorial arguing that a vaccine is the only thing that has ever been effective in eliminating a disease.[12]

Hoffman had raised $63 million and spent almost all of it. Current funds would keep him going for another twelve months. "So now I've got to make the case again that we've got a business plan and economics that make sense for investors. So I don't have to keep begging for money only from foundations," he said. "There is actually a multibillion-dollar market for travelers and the military. And a vaccine can be a lot cheaper than taking Malerone (the current antimalarial of choice) for extended periods."

Some people are as energized by defeat as by victory. Their competitive juices flow stronger when they've picked themselves up off the mat than before they were knocked down. Depending on how long they were down and out, whether they crawled away or were carried, or walked off on their own power, they may be wiser as well. I could sense that Hoffman had bounced back, was undeterred, but I wasn't prepared for what he said next: "So we're going to go forward and do a trial in which the vaccine is administered intravenously. Probably in about four months if we can raise the millions we need. If you can draw blood from someone, as doctors do in offices around the world every day, you can immunize them through an IV inoculation. There is no difference. We're going to prove that the vaccine works and we're going to prove that people can be immunized with IV's."

It would be hard to imagine that the creator of what many consider the most impractical vaccine ever manufactured could actually make its delivery equally impractical, but Hoffman may do just that by using the IV delivery route. Hoffman is not only undeterred, but more enthusiastic than ever. Because while he's not sure how to administer the vaccine, he knows it's still the only one that provides high levels of protection, the only one that can save the lives of a million kids a year and be used to eliminate *Plasmodium falciparum*. The clinical trial helped take at least one option off the table. In the spirit of Thomas Edison, who once said, "I have not failed. I've just found 10,000 ways that won't work," for Steve that means one of the remaining possibilities is all the more likely to be the right one.[13]

CHAPTER 11

PHILANTHROPY'S SHIFTING TIDES

A malaria vaccine has protected a significant percentage of children against uncomplicated malaria, infection, and even severe forms of the disease for at least six months, according to a proof-of-concept study published today in The Lancet. . . .

"Our results demonstrate the feasibility of developing an efficacious vaccine against malaria," wrote Dr. Pedro Alonso, . . . director of the Center for International Health of the Hospital Clinic in Barcelona, Spain.

—Barbara K. Hecht, Ph.D., and Frederick Hecht, M.D.,
"Malaria Vaccine Battle Has Been WON,"
MedicineNet.com, October 14, 2004

THE ECONOMIST SAID IN 2007 that "there is much talk nowadays of a new golden age of philanthropy dawning."[1] It is an era that has indeed been marked by a huge infusion of capital, thanks to economic growth and a technology industry that made many tech-savvy entrepreneurs wealthy at a young age. A greater intergenerational transfer of wealth is anticipated than at any time in our nation's history. It is projected to reach $41 trillion by 2052, according to the Center on Wealth and Philanthropy at Boston College. Even in the face of the recession of 2009–2010, there is excitement about philanthropy's potential, not just because of the amount of money involved but because of new thinking about how it should be deployed.

Indeed, it is an era of tremendous intellectual ferment about the role of philanthropy and how it can have the greatest impact, an era accompanied by experimentation, with each new major philanthropy testing out its own signature style. Organizations such as the NonProfit Finance Fund, which provides growth capital strategies to nonprofits; Venture Philanthropy Partners, which uses venture capital practices in philanthropic investing; and Community Wealth Ventures, which designs earned-income strategies, strive to demonstrate the merits of their approaches.

Perhaps what has changed most significantly is philanthropy's ambition. In philanthropy's earliest days, the wealthy felt an obligation to "give something back," and charitable organizations were the vehicle that enabled them to do so. As social consciousness spread and philanthropy became more grassroots, many simply wanted to genuinely contribute to the aid of others, ensuring that the poor were clothed and fed, that widows and orphans were cared for rather than neglected, that the homeless were sheltered. Others returned wealth to the organizations from which they had personally benefited, such as universities or cultural institutions.

But over the past twenty-five years, and especially in the past ten, the ambition of much philanthropy has shifted from seeing "doing good" as good enough to instead wanting to actually solve complex social problems. It is now not uncommon to have a foundation fund the planning process for solving a problem and to then, in effect, subcontract out the necessary actions to nonprofit grant partners. The Gates Foundation very much acts like the general contractor responsible for eradicating malaria, using a wide variety of subcontractors who specialize in vaccines, drugs, diagnostic techniques, and public health systems. On a smaller scale, the Eli and Edythe Broad Foundation aspires to do the same in transforming public education, as does George Soros's Open Society Institute when it comes to strengthening civil society.

Once the ambition of such institutions changes, the way they allocate funds changes. They now must become long-term institution builders, investing not only in programs, but in operations, capacity, and leadership. They must place greater priority on sustainability and scale. Above all, they must invest not only in creating solutions, but in creating markets that will embrace and distribute their solutions.

This has become the special province of today's most far-reaching social entrepreneurs. If ever there were a label made catchy by the business-worshipping zeitgeist that has held a grip on the nonprofit sector, "social entrepreneurship" is it. If ever there were a label in need of a useful definition, this is it, too.

Nearly everyone engaged in nonprofit work today wishes to share in the aura of social entrepreneurship being touted by philanthropic insiders and repeated in the media. It sounds more impressive, implies more talent and strategy, and personifies more attractive qualities than mere charity work. But not all new nonprofits are entrepreneurial. Being entrepreneurial means more than just being new, or being started by young people, or leveraging private resources for a public purpose, or even starting with a little and turning it into a lot.

What is new—and what excites me—about social entrepreneurship is the determination to find or create markets to enable nonprofit goods and services to get to scale and sustain themselves, instead of having to depend on constant philanthropic subsidization. This means creating commercial markets. Nowhere is this strategy being pursued more vigorously or more purposefully than in the field of global health, especially around the effort to battle the so-called "neglected diseases." But there are examples from other fields as well. Perhaps none is so visible currently as the market for carbon offsets created by companies that cannot cut their own carbon emissions enough to meet legal requirement and so instead pay other companies that have made an excess of cuts, making an offsetting purchase.

THE EMERGENCE OF
MARKET-DIRECTED ORGANIZATIONS

The laws of economics govern global health outcomes every bit as much as principles of biology and sequences of genes do.

Until recently, there has been very little economic interest at stake in addressing malaria and the neglected diseases. So, in the absence of obvious commercial markets, today's entrepreneurs have sought to create their own, or at least have begun to believe that there has been a market hiding in plain sight all along. Steve Hoffman, for example, sees the armed services and American and European travelers as a $3 billion market for the Sanaria vaccine. It is a market large enough to reward a pharmaceutical corporation for investing in a product that could also be made available to children in developing countries for no profit, but ideally at cost.

Bill Gates, at the 2008 World Economic Forum in Davos, Switzerland, observed:

> The great advances in the world have often aggravated the inequities in the world. The least needy see the most improvement, and the most needy get the least—in particular the billion people who live on less than a dollar a day. . . .
>
> Why do people benefit in inverse proportion to their need? Well, market incentives make that happen.
>
> In a system of capitalism, as people's wealth rises, the financial incentive to serve them rises. As their wealth falls, the financial incentive to serve them falls, until it becomes

zero. We have to find a way to make the aspects of capitalism that serve wealthier people serve poorer people as well.[2]

The challenge is twofold. One is whether nonprofits can not only develop a needed product or service, but also have enough sophistication and skill to attract market forces, bring the product or service to scale, and sustain those efforts in order to reach all of those in need. The other is whether philanthropic institutions will be able to provide the risk capital essential to gaining meaningful access to markets in the first place. Increasingly, entrepreneurial funds are doing just that—ranging from the Edna McConnell Clark Foundation, which evolved from a more traditional grant maker, to SeaChange Capital Partners, created in 2006 by two former Goldman Sachs investment bankers expressly for this purpose.

Such an investment approach requires something different from support for entrepreneurship—including social entrepreneurship—and something distinct as well from the passion, innovation, and risk-taking that the word "entrepreneurship" implies. In addition, it requires nurturing a new kind of nonprofit that strategically repositions itself and directs its energies toward long-term market solutions: an NGO, or nongovernmental organization, that transforms itself into an MDO, a market-directed organization.

At some point, most industries or sectors of society are convulsed by genuine revolutions in thinking that transform their growth from the incremental to the exponential.

Manufacturing had its industrial revolution. Technology saw the creation of the silicon chip. Science is currently racing to catch up to the mapping of the genome. Such revolutions have more to do with leaps of imagination than they do with money.

But the nonprofit sector has yet to experience such a revolution. That may be about to change, and bringing market economic principles into social policy may mark the beginning of the shift. It could enable the sector to go beyond good intentions to actually solving and eradicating neglected social diseases in the same way that the neglected tropical diseases are being attacked.

If one believes there is a moral obligation to share our strengths and intellectual gifts to develop solutions to human need, then the moral obligation may be even greater to ensure that such solutions are accessible to everyone, and not just to the privileged few.

As society continues to struggle to strike the right balance between political, nonprofit, and market solutions, philanthropic organizations have a huge opportunity to shape that debate. By taking the kinds of risks philanthropic institutions alone have the ability to take, and helping to manage the risk of nonprofits willing to hold themselves accountable to market forces, the philanthropic sector can provide leadership that can determine the fate of our generation and those of the future.

The new philanthropy, as represented by the Gates Foundation and others, does not supplant what government or the economic marketplace can do. And unlike traditional philanthropy, it is a by-product not only of enormous riches, generated by innovation and acumen in business, but also of shifting political tides.

Andrew Carnegie and John D. Rockefeller operated before the era we think of as big government. It was before Lyndon Johnson's Great Society, before World War II, and even before Franklin Roosevelt's New Deal. Originally, philanthropic funding filled gaps left by political institutions so young that they did not yet fully recognize issues such as literacy and education as a national responsibility.

Bill Gates came along at the opposite end of the spectrum, after the expansion of government and after the backlash that led President Bill Clinton to proclaim that "the era of Big Government is over." Instead of filling the void left by government, like Carnegie, Gates designed funding to bridge gaps left by economic markets, such as the lack of vaccines for neglected tropical diseases, or the shortcomings in efforts to increase agricultural productivity and crop yields in Africa.

With social entrepreneurship principles playing a larger role, philanthropy has evolved to take on tougher, more complex, seemingly intractable problems—problems that other institutions have avoided or abandoned. The philanthropic leap to doing things like attempting to invent vaccines that no one has ever been able to invent before means taking big risks; being willing, at times, to pay for a failure; and committing to long-term investments that may not yield short-term rewards.

This doesn't make the new philanthropy of Bill Gates better or worse than the old philanthropy of Carnegie and Rockefeller. It just makes it a better match for our times.

LESSONS LEARNED:
SIX PROBLEM-SOLVING STRATEGIES

In Washington and the other centers of government that we look to for solutions to social problems, the battle over how to solve our toughest and most controversial problems usually revolves around spending more money or less money, government taking a bigger or smaller role, and the right and left poles of the culture wars. But from today's unprecedented convergence of science, entrepreneurship, and philanthropy, there is emerging a set of problem-solving strategies that, while not apolitical, are certainly less political, and until now have been widely overlooked.

Those from whom I've learned have made no grandiose claims for their work. Their energies have been naturally and appropriately focused on solving the specific and very pressing problems in front of them. They haven't had the luxury of looking across the philanthropic field and discerning patterns that may be useful beyond their own office or lab, or extracting broader lessons.

What follows is a summary of six major lessons I have learned about solving seemingly intractable problems and how this new philanthropy, at its best, can work. There were other small discoveries to be made, of course; but these are the ones that I felt had the most import and utility for others working in a wide range of organizations to bring about change. Although I learned them by observing how leaders in global health seek to solve problems that affect people who are not served by the markets that the rest of us rely upon, their usefulness cuts across many fields and disciplines of social change.

Lesson One: Invest in Bringing Existing Solutions to Scale Rather Than in Discovering New Ones

"The key to developing a malaria vaccine is biotech engineering, not scientific discovery," Steve Hoffman said the first time we met, intentionally overstating the point to make it, because targeted scientific discovery clearly played a critical role in his work. When all else failed, or at least had not succeeded, Hoffman devoted his energies to the daunting engineering required to bring viability to the one discovery that had worked, however impractically, already. Jay Keasling employed a very different methodology to do the same thing with artemisinin. The same approach could be used to solve many other social problems. What we need are not necessarily new solutions, but strategies for making existing solutions affordable, scalable, and sustainable. Until recently, there has been little institutional support, in the form of money, talent, or intellectual capital, for such strategies.

Most problems—whether they have to do with hunger, illiteracy, health care, or education—have a solution, but few solutions have a strategy for getting to scale and being sustainable. Jeffrey Bradach, a former Harvard Business School professor who started the consulting firm Bridgespan to advise nonprofits, put it this way:

> Homelessness, illiteracy, chronic unemployment: nonprofits struggle to address society's most intractable problems. And yet, as Bill Clinton noted, in reviewing school reform initiatives during his presidency, "Nearly every problem has been solved by someone, somewhere." The frustration is that "we can't seem to replicate [those solutions] anywhere else."
>
> With a few exceptions, the nonprofit sector in the United States is comprised of cottage enterprises—thousands upon thousands of programs, each operating in a single neighborhood, in a single city or town. Often, this may be the most appropriate form of organization, but in some—perhaps many—cases, it represents a substantial loss to society overall. Time, funds, and imagination are poured into new programs that at best reinvent the wheel, while the potential of programs that have already proven their effectiveness remains sadly underdeveloped.
>
> . . . Add in the fact that for many social entrepreneurs, autonomy is an important form of psychic income, and it becomes easy to understand why implementing someone else's dream tends not to be nearly as satisfying as building one's own.[3]

In Washington, D.C., a nonprofit called College Summit faces just such a challenge of bringing a proven strategy to scale on a national level. The organization has exceeded all expectations in helping talented, low-income high-school students successfully navigate the college-access process. But its growth has been constrained by lack of funding and low visibility. Mary's Center, in the Adams Morgan neighborhood of D.C., knows how to provide quality maternal and child health care for immigrant Latino families and has successfully reversed horrible infant mortality statistics. If it grew into a national organization, tens of thousands of lives would likely be saved. But after twenty years it has expanded no further than neighboring Maryland. Because the nonprofit sector lacks mechanisms to steer capital toward high-performing organizations and away from the low-performing ones, they often remain unknown to others—and undercapitalized.

One reason this is so is that new ideas, not old ones, have been the darling of traditional philanthropy, and all of the incentives align with pilot programs and starting something, as opposed to scaling something. But that attitude is starting to change.

Chuck Harris, a Goldman Sachs banker for twenty-three years, is the founder of SeaChange Capital Partners, which hopes to fill the market gap that exists for growth capital for nonprofits with proven track records. SeaChange arranges funding for nonprofits that is large enough to be transformational. Harris, who first noticed the problem after becoming a donor to College Summit in 2004, told me: "I decided to work for six months in the development department. I observed the disconnect between the founder, J. B. Schramm, a dynamic entrepreneur with very lofty ambitions, and the way it was being financed, which was by grant writers cranking out as many grants each week as they could. And I thought, why not finance this as a corporation would finance this at the same stage of its growth?"

Much of the entrepreneurship in social entrepreneurship today is finally beginning to focus on scaling, not just creating. They require two different skill sets, and they are rarely found in the same individual. Share Our Strength board member Wally Doolin, who was CEO of the restaurant company that owned TGI Fridays and also served as CEO of casual dining chains La Madeleine and Buca di Beppo, made this very point to me once, explaining, "I could never build a really good restaurant. But I could build one hundred of them." In other words, he was not a creator, but a scaler. Once proof of concept is achieved for a creative vision, Wally uses market forces—consumer demand, investment capital, and so on—to bring it to scale.

Lesson Two: Most Failures in Life Are Failures of Imagination

Most failures are not failures of planning, strategy, resources, organization, or discipline, but failures of imagination. There is always the temptation to point the finger of blame outward. "If only someone would give us more money." "If only we'd been given more time." "If only the consultants would deliver what they promised." The necessary resources always seem just out of reach. While the constraints are real, they are often just symptoms of a more fundamental issue: timid adherence to conventional wisdom. Seemingly intractable problems get solved when someone looks at them in a new way, challenges the very premise of the problem itself, and creates a solution that, until it exists, could not even have been imagined.

Having devoted his entire career to other attempts at creating a malaria vaccine that proved unsuccessful, Hoffman was fully aware of how impractical a whole-parasite vaccine would be. Even nearly a decade after proposing it at a major conference of malaria experts, he recalled what happened when he said, "What about making it into a vaccine?": "You could have heard a pin drop. Not one person thought that was a reasonable thing to do at all. They thought it was ridiculous."[4] But Hoffman never assumed it was ridiculous just because everyone else in the room did. Instead, he persisted, and he has spent years of his life trying to find out if it's possible.

For decades, global health activists sought to cajole large pharmaceutical companies into investing in drugs for neglected tropical diseases. They were mostly unsuccessful. Then Victoria Hale conceived an idea that had simply not been previously imagined: a nonprofit pharmaceutical company, which became the Institute for OneWorld Health.

Or take Teach for America. What could be more impractical than founder Wendy Kopp's vision of placing lightly trained college graduates into some of the toughest urban schools in America, and having them teach poorly prepared teenagers? Impractical? Arguably so. Impossible? Hardly. Wendy's vision allowed her to bridge the imagination gap. In 2007, there were twenty colleges around the country that had 10 percent of their graduating class applying for Teach for America, and ninety colleges that had 5 percent of their graduates applying. With more than 3,000 corps members, the organization ranked number seven on Bloomberg BusinessWeek's 2009 list of "Best Places to Launch a Career."[5]

Leaps of imagination are something quite different from the qualities of creativity that are often desirable in problem solving. Creativity implies generating new ideas or making new associations between old ideas and concepts. But leaps of imagination are not so much about new ideas as about a new conviction of what is possible.

Lesson Three: It's the Economics, Stupid

Agents of change tend to be creative designers more than engineers. They are able to invent solutions, but not always able to make the solutions economical enough to be sustainable. They leave that job to others, usually government. But in the last quarter of the twentieth century, government, lacking sufficient resources, stopped playing that role. Typically those who start or lead an organization to pursue some sort of change end up retrofitting it, belatedly, to include economic considerations. But this is like building a beautifully designed car without installing a gas tank or fuel line. It will go only so far as it can coast.

It is incumbent upon social entrepreneurs to not only develop solutions but to make them affordable. As Dr. Peter Hotez of George Washington University said of the hookworm vaccine he is trying to develop, "If I can't make it for a dollar a dose, I might as well not make it at all." It may be a lot to ask of molecular biologists, tropical medicine physicians, and nonprofit executives to bring not only their creative ideas to their work but to also bring the skills, experience, and wisdom of MBA's, and yet, failing to do so dooms their chances of success.

This point goes to the heart of the nonprofit culture because a prerequisite to scaling an enterprise is investing in the enterprise itself, not just its programs. It's what venture capitalists do to help start-ups—invest not only in their products but in their capacity to produce the products along the lines of an economic model that is sustainable. But this is rarely the way nonprofits and their funders behave.

Clara Miller is executive director of the Nonprofit Finance Fund, which helps nonprofit executives understand the tools of commerce and has lent more than $160 million to organizations for capacity building. One of the clearest thinkers in the country about how nonprofits can use financial strategies to grow, sustain themselves, and increase impact, Miller distinguishes between an organization's programs and the health of what she calls the overall "enterprise." "Philanthropy tends to be enterprise blind and therefore enterprise unfriendly," she told me. "We want to create social value everywhere all of the time. But we have no equity investors, no one to put their arms around the enterprise and protect it." Everyone involved in a nonprofit puts those they serve ahead of the sustainability, even the very survival, of the organization itself. It is noble in the short term, but almost a guarantee that the nonprofit won't be effective in the long-term.

Miller went onto explain, "The path to scale invariably involves profitability. But the challenge in the nonprofit world is that the competition is typically for subsidies, and the bigger you get the worse off you are economically. The enterprise gets exploited again and again in favor of current services. The solution is not necessarily even about capital. It is about *full cost recovery pricing*. We need to embrace the tools of commerce without fooling ourselves that the economic propositions are the same."

Many of the organizations that have been mentioned in these pages—College Summit, Share Our Strength, Teach for America—find themselves admired for their core ideas, but challenged to get to scale. All of them need to explore the economics that will enable them to find their version of a product or service that can be delivered for "a dollar a dose."

Lesson Four: Creating New
Markets or Proxies for Markets

When markets don't exist, they must somehow be created; or at least proxies or substitutes for market forces must be employed as an alternative. "Market" is in some ways just a euphemism for the wider public support that must be marshaled to amass the resources necessary for solving big problems.

Philanthropy itself is a response to the failure of market forces, and within philanthropy proxies for market forces can take many forms. They represent most of the major philanthropic innovations of the past fifteen years, and many fall under the rubric of social entrepreneurship.

Venture philanthropy, a form of grant making or philanthropic investing that adopts venture capital investing strategies and "equity-like" investments for building nonprofit capacity, is one form of proxy for market forces. It often involves long-term commitments of deep financial investment tied to the acceptance of strategic management assistance. For example, when Venture Philanthropy Partners in the greater Washington region invests in a youth-serving organization such as The Seed School, which is an urban, public boarding school, they take a seat on the board, set milestones that will trigger future funding, and provide both cash and expertise in hiring, management, and growth decisions.

Community wealth creation, in which nonprofit organizations actually create new wealth through earned income streams, rather than redistributing charitable wealth, is another example. Sometimes this means leveraging an organization's assets into a business venture whose profits will be devoted to subsidizing and extending the reach of the nonprofit's mission. The Cuyahoga Valley National Park Association, for example, has begun doing facility rentals for social events. It's one way of leveraging the organization's assets to generate incremental revenues that can be used to support and expand the organization's mission in preserving the parks.

Or it may mean the launch of a cause-related marketing campaign in which a corporation and its consumers share the cost and credit that goes with earmarking a percentage of the sale of an item. During the Charge Against Hunger campaign, American Express donated two cents to Share Our Strength every time an American Express card was used for any type of purchase, and the result was $22 million to expand Share Our Strength's anti-hunger activities. Other times this strategy takes the form of fee-for-service programs aimed at segments of the market that can afford to pay. This is what happened when the Jewish Social Service Agency of Rockville, Maryland, began providing home health care to high-income individuals to generate profit that could be used to increase the amount of home health care they provided for the low-income families that were part of their critical mission.

Sometimes a secondary market can be used to subsidize the primary market, which is how Hoffman thinks about the way travelers and military personnel would become customers for his vaccine, helping to generate the revenue needed for his primary focus: the African children malaria is decimating.

Lesson Five: Solving, Not Salving

The solutions that garner the most support, which in turn gives them a better chance of success, are often those that are biggest and boldest and embody the promise of solving a problem once and for all, not just ameliorating it by placing Band-Aids on its symptoms. Finding such a solution requires walking the fine line that the writer and social activist Jonathan Kozol suggested when he advised, "Pick battles big enough to matter but small enough to win."[6]

We learned the same lesson at Share Our Strength when we changed our goal from feeding children to ending childhood hunger. The former, feeding children, is both simple and satisfying, and consequently very popular. It is where most funders want to put their money and where most of their traditional grant recipients want to spend it. The knowhow exists for doing it, and the attendant risks are low. But the kind of emergency food assistance that is offered at soup kitchens and shelters, while necessary, does not treat the underlying problem. And focusing on it exclusively would virtually guarantee that childhood hunger would continue to exist in America.

On the other hand, ending childhood hunger, so that children won't need emergency food assistance in the first place, holds the promise of long-term reward. It also lacks immediate gratification, because it will take at least five years to accomplish, possibly ten. But once we set it as our goal—once we shot down the notion that it was just not possible—it proved ambitious enough to inspire both current and potential future supporters, yet realistic enough to be achieved. Share Our Strength began to grow faster than it ever had before.

Melinda Gates made the same challenge when, in October 2007, at the three-day forum in Seattle, she demanded that malaria not be just managed, but eradicated. She said: "Any goal short of eradicating malaria is accepting malaria; it's making peace with malaria; it's rich countries saying, 'We don't need to eliminate malaria around the world as long as we've eliminated malaria in our own countries.' That's just unacceptable."[7]

There is greater risk of failure when the bar is set so high, and certainly the likelihood of greater expense and complexity. But the reward is a solution worth striving for. Such solutions inspire us. They attract new talent and new resources. They justify the enormously hard work that is required to make a difference.

Though it is not often openly acknowledged, and though nonprofits don't usually behave competitively, the many good organizations and initiatives aimed at addressing social problems do in fact compete directly for the attention and resources of key stakeholders, whether defined as staff talent, financial supporters, policymakers, or all of the above. The pool of philanthropic dollars at any given point in time is finite. Increasingly, knowledgeable and sophisticated donors look for "return on investment" even though it cannot be measured as precisely as it can in the world of for-profit business investments. What return could be better than eradicating a problem once and for all? That's one of the factors that makes solving a problem, rather than improving or ameliorating one, so compelling.

Lesson Six: The Soul of a Competitor

In the social sector, everyone calls for collaboration, which of course has its merits. But competition, at its best, has its merits, too, and what's needed in the nonprofit sector even more than collaboration is a commitment to compete. It is only by competing that nonprofits can avail themselves of the advantages of competition.

I am on the board of the Timberland Company, which makes boots as well as other footwear and apparel. It is my first corporate board experience. I've been on five or six nonprofit boards, but the focus of the Timberland directors is quite different. At Timberland the board is intently focused on whether the company is going to out-perform and out-compete its peers. Board members look at the talent within the company, how it is organized, what products it is making, and other matters through this lens. The most important thing I've learned there is that to compete at any level you must compete at every level.

At Timberland they don't just see themselves as competing to get consumers to buy their footwear. They see themselves as competing to get the best people in the footwear and apparel industries to work at Timberland and to stay there longer than they do at rival companies. Timberland's human resources department measures how long Timberland retains its top talent compared to competitors like Nike, Ecco, Reebok, and Merrill. Likewise for their investments in corporate social responsibility, brand building, and the like. To compete at any level, you must be competitive at every level.

Nonprofits have the opposite instinct. They want to do things with volunteers. Not with the best resources, but with the best leftover resources or donated resources. Such choices are based on noble intentions, and going that route may be economical in the short run, but it completely undermines one's long-term competitive strengths.

Another facet of succeeding as a nonprofit is being competitive economically, that is, delivering your product at a price point—again, Peter Hotez's "dollar a dose," or Jay Keasling's attempt to increase supply through synthetic biology so that life-saving artemisinin becomes affordable to those who need it the most.

Finally, to be effective in solving social problems, whether in hunger, education, the environment, or global health, one must compete to provide the best return on investment from a mission point of view. This requires a willingness to hold oneself accountable for measuring outcomes and communicating what you've measured. The business sector does this through stock price and earnings per share. Hoffman, Alonso, and Keasling will do it through clinical trials. The goal of eradication may seem idealistic, but ultimately, it is measurable, and—as was the case with smallpox—the measurement everyone is aiming for is zero. Getting there may take a long time, but it begins with having the courage of a competitor to set a measurable goal.

Finding solutions that advance the public good takes more than the spirit of a saint, it requires the soul of a competitor. Collaboration is critical to the advancement of most scientific objectives, and progress is always built up in the foundations laid by others. Those at the center of the malaria vaccine development story are no exception, but they also approach their work with the competitive zeal of athletes who work to give themselves every possible edge. They don't expect first-rate outcomes with second-rate inputs. They resist the reflex to depend on whatever leftover resources are available rather than seeking out the best resources available. They have a personal drive to win, they are undeterred by short-term setbacks, and they believe in the superiority of their own vision and abilities, even when no one else does.

CHAPTER 12

MORAL IMAGINATION

No great cause is ever lost or ever won. The battle must always be renewed and the creed restated, and the old formulas, once so potent a revelation, become only dim antiquarian echoes. But some things are universal, catholic and undying—the souls of which such formulas are the broken gleams. These do not age or pass out of fashion, for they symbolize eternal things. They are the guardians of the freedom of the human spirit, the proof of what our mortal frailty can achieve.

—John Buchan, *Montrose: A History*

THE THIRD ANNUAL WORLD Malaria Day, celebrated on April 25, 2010, marked a wave of optimism about the progress that had been made in reducing deaths from malaria. "A wind of change has stirred Africa and the world," declared the executive director of Roll Back Malaria, Professor Awa Marie Coll-Seck. United Nations Special Envoy Ray Chambers, writing in the *Financial Times* about the promise of insecticide-treated bed nets, proclaimed: "Today, we know we can achieve the goal of universal coverage of nets by the end of this year and near zero deaths from the disease by 2015."[1]

Such optimism would have been unthinkable on the day five years earlier when I'd first walked into a cramped, make-shift lab in a Rockville, Maryland, strip mall to learn more about that "wild thing Steve Hoffman is doing." Back then, deaths from malaria were steadily increasing. Bill Gates's Grand Challenges Exploration Grants had yet to award a single dollar. There was little support for innovation or entrepreneurship in global health, and no such thing as a UN special envoy for Malaria.

The battle against malaria was being waged by a relative handful of obscure experts and specialists in research labs, military hospitals, and schools of public health, just as it had been for more than a hundred years. It was the classic "neglected disease"—not on the radar screen of most of the developed world or receiving much support. There were about seventy different vaccines in development, most underfunded, and none instilling confidence.

Now, less than a decade later, Hoffman was at the beginning of his clinical trials, and GlaxoSmithKline's researchers were near the end of theirs. Bill and Melinda Gates had declared it the decade of vaccines, and billions of dollars in new funding had been committed. Jay Keasling had contracted with Sanofi-Aventis to mass-produce artemisinin. The Global Fund to Fight AIDS, Tuberculosis and Malaria was reporting the distribution of 108 million bed nets a year. There had been striking increases in public awareness, political will, and consequently, resources, ranging from the World Bank's increased financial commitment of $200 million to *Idol Gives Back*, the Emmy Award–winning television show on Fox that engaged celebrities to help raise $45 million for the cause from viewers.

Most important of all, there had been at least some measurable results, particularly from control strategies such as insecticide spraying and bed nets. The number of African households with at least one bed net increased from 17 percent to 31 percent between 2006 and 2008, and 9 of 45 malaria-endemic countries—including Eritrea, Swaziland, Botswana, Equatorial Guinea, and Zambia—had seen a 50 percent drop in cases, thanks to a tenfold increase in funding for malaria control since 2004.[2]

Spurred on by such outcomes, other African countries in 2010 were making unprecedented efforts to achieve universal coverage, hoping to cut malaria mortality in half. For example, Congo and Nigeria, which together accounted for 36 percent of the malaria burden, were mounting the largest net-distribution campaigns in the history of malaria control.[3]

No one was debating that a sense of optimism had taken hold. Whether such optimism was warranted, however, remained an open and energetically debated question. In 2009, the World Health Organization reported 243 million cases of malaria and 863,000 deaths attributed to the disease.[4]

As a result, not all global health leaders joined in the celebration. Those with long memories had seen misplaced optimism before, sometimes with devastating consequences. Some saw the danger of becoming a victim of one's own success looming, especially because so much of the success came from methods like using bed nets, which reinforce a narrower, not a more expansive, approach. Bed-net advocates ignore or dismiss vaccine development as a distant ideal. Vaccine developers have little interest in, and rarely advocate for, investing in medicines, a strategy they see as shortsighted compared to developing a vaccine to prevent infection in the first place. This, along with reports of emerging resistance to artemisinin and the belief that even current efforts lack sufficient breadth and ambition, have combined to produce dissenting voices.

The medical journal *The Lancet* timed its World Malaria Day editorial to emphasize that "malaria control and elimination via prevention and treatment can only go so far. The risk of serious setbacks is ever present. What is still needed is the only tool that has ever truly conquered any infectious disease: an effective and affordable vaccine. And here, the global malaria community has been too complacent."[5]

On the Friday before World Malaria Day, the *Wall Street Journal* led with a cautionary tale about the perils facing even the most ambitious vaccine strategies. The front-page story was not about the newly energized global effort to eradicate malaria, but instead about the decades-long struggle to conquer polio and the setbacks that organizations working in the field of global health, including the Gates Foundation, had encountered in their strategy of massive vaccination campaigns.[6] The painful lessons learned may be invaluable to how we think about eradicating not only polio and malaria, but hunger and other persistent challenges as well.

The *Wall Street Journal* article documented new polio outbreaks in a number of African countries—Uganda, Mali, Ghana, Kenya—that had been believed to have stopped the disease. Over the past two decades, $8.2 billion has been spent to kill off polio, with the hope that it would soon be eradicated just as smallpox was in 1979. Bill Gates alone spent $700 million to fight polio. Success seemed close. The number of polio cases in 1988, 350,000, decreased to under 1,000 by the year 2000. But in 2009 new outbreaks brought the total back to 1,600 cases. Once polio was ended in some countries, weak health-care systems, poor sanitation, and malnutrition allowed it to return.

The *Wall Street Journal*, describing it as "one of the most controversial debates in global health," framed the debate this way: "Is humanity better served by waging wars on individual diseases, like polio? Or is it better to pursue a broader set of health goals simultaneously—improving hygiene, expanding immunizations, providing clean drinking water—that don't eliminate any one disease, but might improve the overall health of people in developing countries?"[7]

Big donors usually prefer the "vertical" strategy of fighting individual diseases, and advocates are influenced by the preferences of big donors. When a more "horizontal" strategy is pursued, focusing on a broader spectrum of health-related issues, the objectives are often less specifically defined. Deadlines are often nonexistent, since improvements are likely to take many years longer than in vertical-style projects. And yet the horizontal strategies can be just as important as the vertical ones to long-term success. The Gates Foundation and allied organizations, the *Wall Street Journal* said, were devising a revamped plan that would represent a major rethinking of strategy acknowledging "that disease-specific wars can succeed only if they also strengthen the overall health systems in poor countries."[8]

It seems an obvious and commonsense insight, but it reverses decades of conventional wisdom that most of the global health community embraced, funded, and acted upon. It also underscores that very different ways of thinking and very different strategies can not only complement each other but be necessary if diseases like malaria and polio are to be drastically reduced or eradicated.

How could Gates and his colleagues have missed this before now? It's not that they aren't smart. Resources were not a primary constraint. Gates has an impressive track record of matching his big bank account with big ideas. No one has ever accused Microsoft's founder of lacking vision or thinking small. But even Gates suffered a failure of imagination when it came to fighting polio. The enormous financial commitment he made to the disease-specific approach must have seemed like a leap of imagination in and of itself, perhaps the bolder course in the either/or choice described above. But bolder still is the decision to do both, notwithstanding the pressure it creates to generate needed resources, maintain focus, and ensure the ability to measure results. It shows that those fighting diseases as intractable as polio or malaria—or taking on any other task of that size—have to ask whether even their most ambitious efforts lack vision and imagination.

WHEN MAN AND MOMENT INTERSECT

Against the backdrop of the millions of lives threatened by the emerging resistance to artemisinin, and a malaria community energized by new funding to support a wide variety of creative experiments, stands the curious figure of Steve Hoffman.

Doggedly championing what is perhaps the most unconventional approach of all, Hoffman had taken on and refuted every argument—purity, safety, effectiveness, stability, feasibility—for why his vaccine could not work. When he was done, there was nothing left but an immunogen that had demonstrated more than 90 percent effectiveness, and that the FDA had approved for Phase I clinical trials as a vaccine.

Why Hoffman? What propelled him and his idea past literally thousands of others? What has brought him to the brink of success?

As is often the case, the answer lies at the intersection of the man and the moment, a combination of his personal qualities and the trends and the times in which he lives.

Hoffman's personal qualities can be summarized in three words: imagination, entrepreneurship, and leadership. I've never heard Hoffman speak in such terms; he only demonstrates and embodies them. The combination of all three is occasionally seen in business or politics, but rarely in science. In Hoffman's hands, it has proved a potent formula.

Hoffman's hard-earned technical successes transformed the perception of his vaccine from preposterous to miraculous. More important, he had the imagination and vision to see each scientific and technological breakthrough not as an end in itself but as the means to a larger end. The time he devoted to science was more than matched by the time he spent coaxing the scientific and philanthropic community up and over Mount Improbable, that Everest-like mountain of skepticism that had prevented them from seeing a potential solution lying long dormant, but nevertheless in front of them all along.

Art instructors urge aspiring artists not to paint what they think they see or what their minds have conditioned them to see, but what they actually see in front of them. Hoffman championed a similar discipline: peeling away preconceived notions about the drawbacks of a vaccine based on live, attenuated parasites and insisting instead on a notion at the heart of all good science—that the vaccine be judged on the evidence and facts. And like Brother Thomas Bezanson, who never hesitated to break a pot that was good but not good enough, Hoffman steadfastly rejected RTS,S, which he'd even had a hand in creating, and any other vaccine that fell short of the gold standard of nearly 100 percent protection.

The technical barriers to success were not inconsiderable. When the yield of parasites harvested per mosquito was not to scale, Hoffman and his team figured out how to increase it. When the time came to irradiate the mosquitoes so that each one received the same dose of radiation, he figured out how to do that, too. Hoffman's wife, Kim Lee, had solved issues of purity and sterility. On and on it went, through years of tedious testing, revision, and refinement.

Over a decade there were hundreds of such technical challenges. They related to how the vaccine would be created, handled, stored, transported, and delivered, at what temperature, in what material, in compliance with which regulations, and so on. And still there are challenges. How will such a vaccine, once manufactured, actually get to Africa and into the infants and children under five years old, the group that needs it most?

Skepticism remained, but each technical achievement became a foothold on Mount Improbable, bringing its glittering peak more clearly into view. Hoffman had seen the summit in his mind's eye; now he had to drag and cajole the whole climbing team that straggled behind him.

The skepticism was not entirely unwarranted. Hoffman's assault on the malaria parasite was neither obvious, at first, nor practical. It was fraught with hardship and with hurdles so steep as to be considered unimaginable, and if imagined, then so difficult, complex, expensive, and tedious as to be unacceptable. As Hoffman along with several colleagues wrote in the journal *Human Vaccines*, the initial insight "was accompanied by an equally universal consensus that it was inconceivable to consider developing an attenuated *P. falciparum* sporozoite vaccine . . . not due to concern about the potential safety. . . . Rather, it was believed to be impossible to manufacture and administer adequate quantities of aseptic, purified, well-characterized, stable *P. falciparum* sporozoites that met regulatory and cost of goods requirements."[9]

The doubters were simply holding to conventional wisdom. But for Hoffman, conventional wisdom prevailed only until weighed against the alternative: the disaster of death and destruction inherent in maintaining the status quo. Hoffman insisted, repeatedly, that his vaccine be judged that way: not on its own, but in comparison to the alternatives.

Many said the beginning of Hoffman's Phase I clinical trial marked a critical turning point. One tropical disease expert, Michael Good, the director of the Queensland Institute of Medical Research, called the trial a "watershed event" and went on to say it was "the culmination of a remarkable translational research effort by Sanaria." Many had believed the vaccine would not be able to meet the FDA's rigorous requirements for safety, sterility, purity, potency, and reproducibility, but Sanaria, said former president of Merck Vaccines, Adel Mahmoud, had "been able to systematically overcome obstacle after obstacle." Myron Levine from the University of Maryland explained that previous research had never been translated into vaccine development "because the task was considered to be impossible."[10]

These comments underscored a principal ingredient of Hoffman's success. "Remarkable translational research effort" is scientific jargon for relentless entrepreneurship. The classic hallmark of entrepreneurship is a willingness and ability to adjust, evolve, and adapt, along the lines of Darwin's explanation of evolution: It is not the smartest or the fastest or even the strongest that survive, but those most able to adapt.

Hoffman reminds me of one of those robotic floor sweepers or battery-operated kid's toys that, when running into a wall, simply careens off into another direction. There's no sign of it being worse for the wear, no matter how many times it bounces back, and eventually it has covered the entire floor.

Hoffman combines imagination and entrepreneurship with leadership, inspiration, calculation, and strategy, and especially with what the historian Richard Neustadt said characterized effective U.S. presidents, the power to persuade. Neustadt explained that the power to persuade "is more than charm or reasoned argument"; it is enabling others to see why it is in their self-interest to act in a particular way.[11]

Just as science, entrepreneurship, and philanthropy evolved, so, too, had Steve Hoffman—from one man with a vision to the leader of an enterprise that has attracted diverse and idealistic talent from around the world as well as funding and the increasing respect of the scientific establishment.

Hoffman's entire enterprise was built on a slender but tantalizing experiment: In the 1990s, Hoffman and thirteen volunteers, testing Ruth Nussenzweig's 1967 work, allowed themselves to be bitten by irradiated, weakened mosquitoes about 1,000 times to simulate a natural immunity. When later "challenged" by being exposed to and bitten by regular infected mosquitoes, thirteen of the fourteen were protected from malaria infection. There had been no injections, no lab-created formula or vaccine, only this crude compression and mimicry of nature's own methodology. From this one set of results, from a sample size smaller than a Little League team, and an experiment tried once and never repeated, Hoffman parlayed his power to persuade into tens of millions of dollars, international attention, and ultimately FDA approval for trials.

Hoffman committed himself not just to science but to leadership in a way that embodied the prescription of Warren Bennis, University of Southern California professor of business administration, founding chairman of USC's Leadership Institute, and author of one of the most influential leadership books of all time, *On Becoming a Leader*:

> Limits, constraints and reduced expectations are the conventional prescriptions for our time. True leaders, however, are able to see beyond an anemic zeitgeist in order to sense opportunities that can employ and house a multitude.
>
> Optimists have a sixth sense for possibilities that realists can't or won't see. That gives the optimist the ability to

"define reality" for others in a compelling way—which is the first task of a leader, as the author Max Dupree has observed. This is not sentimentalism: It is the essence of creative pragmatism. It is good because it works.[12]

For Hoffman this has been a way of thinking and a way of being more influential than any breakthrough or "Eureka moment" in the traditional sense. Hoffman didn't have a strategy, or a formula to be replicated. Instead he had a mindset. He came upon the vaccine the way Sherlock Holmes came upon the solution to a mystery: "When you have eliminated the impossible, whatever remains, however improbable, must be the truth." It would be hard to have a vaccine more improbable than Sanaria's, but it is the improbable that Hoffman is on his way to proving true.

Although it will be years before we know the final results of Sanaria's clinical trials with certainty, and the ultimate success or failure in overcoming all of the other technical, logistical, political, and economic hurdles to eradicating malaria will remain uncertain for even longer, we at least know more about the ingredients of breakthrough thinking that bring one to the pinnacle of such success. Not everyone who makes it to the base camp of Mt. Everest makes it the rest of the way to the top. But just getting to the base camp is a Herculean task that separates the very few from the very many rest of us. Knowing how they got that far is no guarantee that one can go the whole way, but it's not a bad place to start.

Hoffman is far from alone there. Rip Ballou, David Lanar, Pedro Alonso, Victoria Hale, and Peter Hotez have all journeyed heroically. At times they've collaborated and at times they've competed, but from each other's successes and failures they have always learned. As a result they have not only advanced their own agendas but also the field of global health, creating hope and inspiring others to tackle problems that affect the most vulnerable and voiceless among us.

WHAT WE SEEK TO KNOW

In May 2010, the United Nations hosted a special photographic exhibit called "Malaria: Blood, Sweat and Tears."[13] The photos were taken in Cambodia, Uganda, and Nigeria. They show people who either have the disease or are engaged in fighting malaria in some way—as a community health worker, a guard at a bed-net warehouse, or a pharmacist. Each picture is compelling on its own, but when you consider the group of photos as a whole, and examine common features, you gain a better understanding of what these African people must endure, and that is what makes this photographer's effort to bear witness so powerful.

I recognized something in one of the women in the photos. She is attractive, perhaps in her late twenties, with jet black hair and high cheek bones shining in the sun. She is wearing a colorful, flowered blouse and carrying her feverish son, with a green towel draped around his shoulders. They are outside, with one of the lush green hillsides of Cambodia behind them, just slightly blurred.

From the way her body is angled it looks as though she may be balancing in the back of a truck. Her son's chin is tucked between her left arm and breast, and her strong left hand presses against his back to steady him as they race toward their destination. His lower jaw is pulled slightly to the left, as if his teeth are chattering from severe chills. His eyelids are heavy, almost closed. But not her eyes. In fact, her eyes burn fiercely, not with fever but with frightened determination.

I've never met this woman, but I recognize her because I can see from her urgency and selflessness that she is every mother I've ever known. Her name is Pheap Sung. She told the photographer that, "He was sick for three days, had a very high fever. I would have sought help at a private clinic, but I did not have the money. The free clinic is a long way, but I decided I had to take him. I thought he might have malaria."[14]

I doubt it would have made any difference if the clinic had been five times as far. There is no such thing as unreasonable when it comes to a mother doing what is necessary for her child. There is no such thing as too far, too much, too expensive, or too complicated. The look on her face was a plea, a look that could go right through the camera's lens all the way across the globe to Steve Hoffman or Rip Ballou or Jay Keasling or Victoria Hale, a plea to not stop at the conventional response, to not be deterred by the unreasonable, to not accept that good is good enough, to not succumb to a failure of imagination.

The 3,000 African kids who die every day from malaria die quietly and invisibly. That's because they die routinely, year in and year out, in numbers too large to fathom. They die in the pages of medical journals, not in our living rooms on high-definition TV. Unlike a child buried in the rubble after Haiti's earthquake, they don't reach the threshold for Anderson Cooper or the 82nd Airborne, or for benefit rock concerts on MTV.

This tension between the immediate and the long-term, between the personal and the abstract, is always with us. It exists in every effort to create meaningful change. The drama of tragedy always prevails over the numbing of statistics. Saving 8 million lives over ten years might get a headline—on a slow news day. The improbable rescue of one child a day from a collapsed building in Port au Prince can lead the news for weeks.

The consequence, while understandable, is a spectacular failure of imagination. When we focus on the one rather than the many, on the symptom rather than the cause, on what we can accomplish on our own rather than on what needs to be accomplished by the broader community, we neglect our greatest opportunities to do the greatest good. It is equivalent to suffering a massive stroke that leaves one seeing only what is in our direct line of sight, with no peripheral vision or sense of relationship to the larger, surrounding world.

There is no recourse to such failure of imagination but to recognize it, confront it, and struggle to overcome it as one might a crippling stutter.

It would be nice if there were a more concrete and guaranteed prescription, perhaps a handy checklist to tick through. But overcoming failures of imagination has less to do with following procedure or tapping external resources than it has to do with looking deeply and expansively within. It requires each of us to intentionally challenge our own imagination, questioning whether we have engaged it to the fullest, and especially pushing to contemplate, and react to, not only what we see but also to what we do not see.

A few years ago, the commencement speaker at a college graduation made exactly this point. The speaker, Ophelia Dahl, cofounder of Partners in Health (PIH) and daughter of the children's book author Roald Dahl, quoted Adam Hochschild, who wrote about the importance of "drawing connections between the near and the distant." Dahl, speaking to the women in the class of 2006 at Wellesley College, went on to explain one of the ingredients most essential to fighting for whatever might be their cause:

> Linking our own lives and fates with those we can't see will, I believe, be the key to a decent and shared future. . . .
>
> Imagination will allow you to make the link between the near of your lives with the distant others and will lead us to realize the plethora of connections between us and the rest of the world, between our lives and that of a Haitian peasant, between us and that of a homeless drug addict, between us and those living without access to clean water or vaccinations or education, and this will surely lead to ways in which

you can influence others and perhaps improve the world along the way.[15]

Dahl said that being the daughter of writer Roald Dahl meant learning a lot about imagination at an early age. She implied that it had served her well in helping to envision and create Partners in Health. After all, PIH had succeeded where so many others had failed precisely because of a leap of imagination. The leap was not that highly educated doctors in Boston would volunteer to provide health care to Haitians in Haiti—though it would be fair to call that a stretch in its own right—but rather that with the support of partners from Boston, Haitians could create and deliver their own health care. *That* is where imagination really triumphed.

The photos at the UN exhibition satisfied Ophelia Dahl's challenge to draw connections and link "our own lives and fates with those we can't see." I recognized Pheap Sung because it was in her face that I also finally met and recognized the mother of Alima, the young Ethiopian schoolgirl I'd befriended, whose photo still graces my bookshelf, who died of cerebral malaria before reaching the age of fourteen, and who should not have. The terror, love, and determination in the eyes of Alima's mother could not have been much different from what I saw in the eyes of Pheap Sung.

We can't all go to Ethiopia, Uganda, or Haiti, or perhaps even to a malaria vaccine lab in a strip mall across town, to witness suffering or to fully understand the need, opportunity, and possibility. But if we are purposeful about using our moral imagination, we shouldn't have to.

In today's world more than at any time in human history, we have access to all of the information we need to bridge the chasm between distant and near. The question is what we will do with it: whether we will not only analyze and categorize and think about it, but also let ourselves feel something about it and act on those feelings.

Compassion is both blessing and balm. But unless it is hitched to the power of imagination, it can leave us one step behind the next tragedy, and the next, always a day late and a dollar short. We'll likely end up doing good, but not nearly good enough.

Moral imagination is supposed to be what differentiates us from other species. But our boast is bigger than our bite. We remain only partially evolved, a work in progress to be admired and resisted both at once. We find ourselves, as Bruce Springsteen sings, "halfway to heaven and just a mile outta hell."

If we hope to truly change the world rather than just the bits and pieces of it that drift in front of us, we must reach for more than the traditional tools stored in those drawers we glibly label "social entrepreneur," "business leader," or "politician." Indeed, we must reach inside, not out; we must shape our own evolution, with faith that the greatest value we can deliver may lie not in what we know but in what we seek to know.

ACKNOWLEDGMENTS

THIS BOOK WAS MADE BETTER by many heads and hands—especially by PublicAffairs editor Clive Priddle, who created order out of chaos. He never lost sight of the story I wanted to tell, even when I kept it pretty well hidden. Thanks also to my initial editor, Morgan Van Vorst, and to Susan Weinberg, publisher, and Peter Osnos, founder and editor-at-large of PublicAffairs, for taking a gamble on this book and for their faithful commitment to giving voice to those who might otherwise be voiceless. Katherine Streck-fus's copyedit improved every page and I deeply appreciate her extraordinary diligence, as well as the formidable production skills of Melissa Raymond. I am also grateful to my longtime friend and agent, Flip Brophy, for helping to make such a perfect match, as she has done many times before.

There were many experts in the global health field who generously guided me through their own stories as well as the history and science of malaria and vaccine development. I have the utmost respect for their expertise and dedication. They include Ruth Nussenzweig, Pedro Alonso, Peter Hotez, David Lanar, Jay Keasling, Kinkead Reiling, Jack Newman, Paul Roepe, Dan Carucci, Marcelo Jacobs-Lorena, Judith Epstein, Victoria Hale, Ray Chambers, and Brian Greenwood, and also Regina Rabinovich, Joe Cerell, and their colleagues on the staff of the Bill and Melinda Gates Foundation.

My greatest debt of gratitude goes to Steve Hoffman and his wife, Kim Lee Sim, who repeatedly and patiently opened up their lab and lives so that I could get a glimpse of the human trials and tribulations behind the science. Their work, assisted by their three sons, has genuinely been a family affair, and if the way Alexander, Ben, and Seth have turned out is any indication of how Sanaria's vaccine to eradicate malaria will fare, the world has reason for hope.

At critical junctures along the way, I was sustained by creative and caring friends, including, first and foremost, Jeff Swartz, as well as Joel Fleishman, Harris Wofford, Chris and Diana Chapman-Walsh, Leah and Bill Steinberg, Rick Russo, David and Katherine Bradley, Sue and Bernie Pucker, and my sister-in-law and babysitter extraordinaire, Patti Jordano.

The staffs of Share Our Strength and Community Wealth Ventures generously took up the slack, as they always have when my attentions were focused elsewhere. Their unrelenting determination to end childhood hunger, and the sacrifices they've made in pursuit of that goal, speak volumes about who they are, the choices they've made, and their shared commitment to those who are the most vulnerable and voiceless among us. Special thanks to my executive team colleagues Pat Nicklin, Chuck Scofield, Eric Schweickert, and Josh Wachs, and also to Amy Celep from Community Wealth Ventures. My sister Debbie, cofounder of Share Our Strength, always an indispensable leader, has once again been a selfless champion of my attempt to put words to the lessons we've learned. Thanks to Alice Pennington for her enthusiasm in keeping me and the manuscript organized, and for always being such a thoughtful colleague and caring friend, and also to Sarah Sandsted for so capably fielding all manner of assignments large and small.

In this and all I do I've been inspired by the strength and resilience of my older son Zach, the determination of my daughter Mollie, and the curiosity and joyfulness of my young niece Sofie Shore.

My wife, Rosemary, has been a partner in this project in every way, with characteristically unfailing instincts upon which I've greatly depended. She and our son, Nate, a fount of imagination in his own right, bore the brunt of my conflicting desires to write and keep my day job. But their love, energy, and spirit made the long hours and absences bearable, and finishing the book especially rewarding. In the course of my research I frequently realized just how many things I don't know, but I do know how lucky I am to have them at the center of my life.

NOTES

CHAPTER 1

1. *The Little Prince*, by Antoine de Saint Exupéry and translated by Katherine Woods, is available online at http://www.angelfire.com/hi/little prince/frames.html. The quotation is from Chapter 21.

2. See Brother Thomas Bezanson's obituary at http://www.icgerie.com/homes/dusckas/obit/2007/08/bezanson.html. See also the page on Brother Thomas at the Pucker Gallery website, http://www.puckergallery.com/bt.html.

3. Rosemary Williams, ed., and Bill Aron, photographer, *Creation Out of Clay: The Ceramic Art and Writings of Brother Thomas* (Grand Rapids, MI: William B. Eerdman's, 1999), 90.

4. Victoria G. Hale, "Our Story: How OneWorld Health Was Founded," http://www.oneworldhealth.org/story.

5. See "Our Impact," Teach for America, http://www.teachforamerica.org/mission/our_impact/our_impact.htm.

6. See Harlem Children's Zone, "The HCZ Project: 100 Blocks, One Bright Future," http://www.hcz.org/about-us/the-hcz-project.

7. Sun Tzu, *The Art of War* (Mineola, NY: Dover Publications, 2002).

CHAPTER 2

1. For biographical material on Steve Hoffman, see the Sanaria website, http://www.sanaria.com/index.php?s=44.php, and Thomas C. Luke and Stephen L. Hoffman, "Rationale and Plans for Developing a Non-Replicating,

287

Metabolically Active, Radiation-Attenuated Plasmodium Falciparum Sporo-zoite Vaccine," *Journal of Experimental Biology* 206 (2003): 3803–3808; first published online September 23, 2003, at http://jeb.biologists.org/cgi/reprint/206/21/3803.

2. Quotes from scientists and others that are not otherwise attributed are from interviews I conducted with them in person or by phone. Often we met more than once, in some cases numerous times over a period of several years. Steve Hoffman and I met regularly over a period of nearly five years at his office or over meals, and we talked frequently by phone and e-mail.

CHAPTER 3

1. Paul R. Russell, "Introduction," in *Preventive Medicine in World War II*, vol. 6, *Communicable Diseases: Malaria*, Col. John Boyd Coates, Jr., ed. (Washington, DC: Medical Department, U.S. Army), available at http://history.amedd.army.mil/booksdocs/wwii/Malaria/chapterI.htm.

2. Committee on U.S. Military Malaria Vaccine Research, *Battling Malaria: Strengthening the U.S. Military Malaria Vaccine Program* (Washington, DC: National Academies Press, 2006), 14. See also Malaria Vaccine Initiative, "Fact Sheet: Malaria and the Military," 2004, http://www.malaria vaccine.org/files/FS_Malaria-Military_9-15-04.pdf.

3. "What Is Malaria?" Malaria Site: All About Malaria, April 14, 2006, http://www.malariasite.com/malaria/WhatIsMalaria.htm.

4. Ethiopian North American Health Professionals, "Facts and Statistics," http://www.enahpa.org/.

5. Dr. Denise Doolan, Patent Application, http://www.faqs.org/patents/app/20080248060; Mark Zottola, "Computational Chemistry, Anti Malaria Drug Research," U.S. Army Research Laboratory, June 3, 2009, http://www.arl.hpc.mil/Publications/eLink_Spring03/malaria.html.

6. Craig Smith and Arthur Hooper, "The Mosquito Can Be More Danger-ous Than the Mortar Round: The Obligations of Command," *Naval War College Review* (Winter 2005), http://www.thefreelibrary.com/The+mosquito+can+be+more+dangerous+than+the+mortar+round%3A+the+-a012 9363033.

7. Ibid.

8. Ibid.

9. Ibid.

10. Ruth Nussenzweig, conversation with the author in her lab at New York University, March 18, 2009; Douglas Birch, "The Struggle to Vanquish an Ancient Foe," *Baltimore Sun*, June 18, 2000.

11. Robert Langreth, "Booster Shot," Forbes.com, November 12, 2007, http://www.forbes.com/free_forbes/2007/1112/078.html?partner =yahoomag.

12. David Biello, "Self-Experimenters: Malaria Vaccine Maven Baits Irradiated Mosquitoes with His Own Arm," *Scientific American*, March 12, 2008, http://www.scientificamerican.com/article.cfm?id=malaria-vaccine-researcher-lets-misquitos-bite-him; Michael Myser, "The Malaria Fighter," *Business 2.0*, January/February 2006.

13. William Deresiewicz, "Solitude and Leadership," AmericanScholar.org, Spring 2010, http://www.theamericanscholar.org/solitude-and-leadership/.

CHAPTER 4

1. The company website is at http://www.sanaria.com/. For a photo of the new headquarters, see http://www.sanaria.com/index.php?s=48.php.

2. Jason Fagone, "The Scientist Ending Malaria with His Army of Mosquitoes," *Esquire*, December 8, 2008, http://www.esquire.com/features/best-and-brightest-2008/malaria-prevention-1208.

3. E. Nardin, F. Zavala, V. Nussenzweig, and R. S. Nussenzweig, "Pre-Erythrocytic Malaria Vaccine: Mechanisms of Protective Immunity and Human Vaccine Trials," *Parassitologia* 41, nos. 1–3 (1999): 397–402.

4. John F. Kennedy, address at Rice University, Houston, Texas, September 12, 1962. See text of speech online at http://www.jfklibrary.org/Historical +Resources/Archives/Reference+Desk/Speeches/JFK/003POF03Space Effort09121962.htm.

CHAPTER 5

1. Donald Burke, American Society of Tropical Medicine and Hygiene Centenniel Address, Philadelphia, December 3, 2003, http://www.astmh .org/AM/Template.cfm?Section=Meeting_Archives&Template=/CM /ContentDisplay.cfm&ContentID=1500.

2. P. Trouiller, P. Olliaro, E. Torreele, J. Orbinski, R. Laing, and N. Ford, "Drug Development for Neglected Diseases: A Deficient Market and a Public-Health Policy Failure," *The Lancet* 359, no. 9324 (2002): 2188–2194.

3. Hélène Delisle, Janet Hatcher Roberts, Michelle Munro, Lori Jones, and Theresa W. Gyorkos, "Review: The Role of NGOs in Global Health Research for Development," *Health Research Policy and Systems* 3, no. 3 (2005), http://www.health-policy-systems.com/content/pdf/1478-4505-3 -3.pdf.

4. Michelle Barry, Presidential Address, 51st annual meeting of the American Society of Tropical Medicine and Hygiene, Denver, Colorado, November 12, 2002, http://www.astmh.org/AM/Template.cfm?Section =History_of_ASTMH&Template=/CM/ContentDisplay.cfm&ContentID =1323.

5. Gary Taubes, Science Watch Newsletter, http://esi-topics.com/malaria/ interviews/StephenHoffman.html, May 2006.

6. B. H. Kean, with Tracy Dahlby, *MD: One Doctor's Adventures Among the Famous and Infamous from the Jungles of Panama to a Park Avenue Practice* (New York: Ballantine Books, 1990).

7. Guy Charmot, "Lavaran and the Discovery of the Malaria Parasite," Centers for Disease Control and Prevention, February 8, 2010, http://www .cdc.gov/malaria/about/history/laveran.html.

8. Dr. Ronald Ross, speech at the Nobel Banquet in Stockholm, December 10, 1902, reprinted in *Nobel Lectures: Physiology or Medicine (1901– 1902)* (Singapore: Published for the Nobel Foundation by World Scientific Publishing Co., 1999).

9. Stephen Hoffman, Presidential Address to American Society of Tropical Medicine and Hygiene, December 2001, *American Journal of Tropical Medicine and Hygiene* 67, no. 1 (2002): 1–7.

10. Karen Hopkin, quoting Dr. Joseph DiRisi, in "SARS, Malaria, and the MicroArray," *The Scientist*, November 21, 2005, http://www.the-scientist .com/article/display/15887/.

CHAPTER 6

1. Philip Bejon, John Lusingo, Ally Olotu, Amanda Leach, Marc Lievens, Johan Vekemans, Salum Mshamu, et al., "Efficacy of RTS,S/AS01E Vaccine Against Malaria in Children 5 to 17 Months of Age," *New England Journal of Medicine* 359, no. 24 (2008): 2521–2532, http://content.nejm.org/ cgi/content/short/359/24/2521.

2. Patrick Adams, "The Sanaria *PfSPZ* Malaria Vaccine, Until Recently Considered Impossible, Is Entering Phase II Trials," TropIKA.net, http:// www.tropika.net/svc/interview/Adams-20091216-Interview-Hoffman.

3. Brian W. Simpson, with photos by Mark Lee, "Putting the Bite on Malaria," *Johns Hopkins Public Health*, Fall 2001, http://www.jhsph.edu/ magazineFall01/Feature1.htm. The authors were summarizing a more in-depth description that is supplied in the classic work by Andrew Spielman and Michael D'Antonio, *Mosquito: A Natural History of Our Most Persistent and Deadly Foe* (New York: Hyperion, 2001).

4. Ethne Barnes, *Diseases and Human Evolution* (Albuquerque: University of New Mexico Press, 2005), 77. I relied on Barnes for much of the description of how the parasite attacks malaria victims.

5. Carole Long graciously allowed me to sit in on a class she teaches at the Uniformed Services University of the Health Sciences, near the National Naval Medical Center in Bethesda, Maryland. The quotation is from that class.

6. A. Ghosh, M. J. Edwards, and M. Jacobs-Lorena, "The Journey of the Malaria Parasite in the Mosquito: Hopes for the New Century," *Parasitology Today* 16, no. 5 (2000): 196–201.

7. U.S. Congress, Office of Technology Assessment, *Pharmaceutical R&D, Costs, Risk and Rewards*, OTA-H-522 (Washington DC: U.S. Government Printing Office, 1993).

CHAPTER 7

1. Lynn Yarris, "*Synthetic Biology Can Help Extend Anti-Malaria Drug Effectiveness,*" News Center, Berkeley Lab, March 3, 2009, http://news center.lbl.gov/feature-stories/2009/03/03/synthetic-biology-can-help -extend-anti-malaria-drug-effectiveness/.

2. Barry Gardner, "Developing Artemisinin," Wellcome Trust website, January 10, 2002, http://malaria.wellcome.ac.uk/doc_WTD023861.html.

3. Donald G. McNeil, "Millions of Lives on the Line in Malaria Battle," *New York Times*, January 25, 2005, http://www.nytimes.com/2005/01/25/ health/policy/25mala.html; Donald G. McNeil, "Deal Seeks to Offer Drug for Malaria at Low Price," *New York Times*, July 18, 2008, http://www .nytimes.com/2008/07/18/health/18malaria.html.

4. Michael Specter, "A Life of Its Own: Where Will Synthetic Biology Lead Us?" *New Yorker*, September 28, 2009, http://www.newyorker.com/ reporting/2009/09/28/090928fa_fact_specter.

5. Andrew Jack, "Novartis Chief in Warning on Cheap Drugs," *Financial Times*, September 30, 2006, http://www.ft.com/cms/s/0/6cfd37e8-5020 -11db-9d85-0000779e2340.html.

6. Keasling has said the same thing in other interviews. For example, see Michael Specter, "A Life of Its Own," *New Yorker*, September 28, 2009.

7. Elizabeth Corcoran, "Stalking a Killer," *California* magazine," November/December 2006, http://alumni.berkeley.edu/news/california -magazine/november-december-2006-life-after-bush/stalking-killer.

8. Robert Sanders, "Keasling and Cal: A Perfect Fit," UCBerkeleyNews, December 13, 2004, http://berkeley.edu/news/media/releases/2004/12/13 _keasling.shtml.

9. Erica Check Hayden, "In the Field," Nature.com, February 13, 2009, http://blogs.nature.com/news/blog/2009/02/aaas_synthetic_biology_races_t .html.

10. Kimberlee Roth, "A Love of Science and a Vision to Save Millions of Lives Make Her Day," *Chronicle of Philanthropy*, April 6, 2006.

11. "Victoria G. Hale," *Nature Reviews Drug Discovery* 4 (November 2005).

12. Ibid.

13. Hale later left OneWorld Health and founded Medicines 360, a non-profit pharmaceutical aiming to address unmet needs of women and children.

14. Peter Hotez, interview with the author, September 13, 2007, and World Health Organization, Institute for Vaccine Research, http://www .who.int/vaccine_research/diseases/soa_parasitic/en/index2.html.

15. "Ending Disease, Ending Poverty: An Interview with Lee Hall and Peter Hotez," America.gov, U.S. Department of State, March 5, 2007, http://www.america.gov/st/scitech-english/2009/April/20090430123720 wrybakcuh0.3490214.html.

16. Lecture at UGA Global Diseases Series, University of Georgia at Athens, February 28, 2006.

17. Merrill Goozner, "Stopping Hookworm," *The Scientist* 21, no. 7 (2007): 52.

18. Peter J. Hotez, David H. Molyneux, Alan Fenwick, Jacob Kumaresan, Sonia Ehrlich Sachs, Jeffrey D. Sachs, and Lorenzo Savioli, "Control of Neglected Tropical Diseases," *New England Journal of Medicine* 357, no. 10 (2007): 1010–1027, http://content.nejm.org/cgi/content/full/357/10/1018.

19. Melinda French Gates, "Malaria Forum Keynote Address," transcript at Bill & Melinda Gates Foundation website, October 17, 2007, http:// www.gatesfoundation.org/speeches-commentary/Pages/melinda-french -gates-2007-malaria-forum.aspx.

20. Ibid.

21. Ibid.

22. "Remarks of Mr. Bill Gates, cofounder of the Bill and Melinda Gates Foundation, at the World Health Assembly," 58th World Health Assembly, Geneva, Switzerland, May 16, 2005, transcript at World Health Organization website, http://www.who.int/mediacentre/events/2005/wha58/gates/ en/index.html.

23. Malaria R&D Alliance, Malaria Research and Development, "An Assessment of Global Investment," November 2005.

24. Ibid.

25. Robert W. Snow, Carlos A. Guerra, Juliette J. Mutheu, and Simon I. Hay, "International Funding for Malaria Control in Relation to Populations

at Risk of Stable *Plasmodium falciparum* Transmission," *PLoS Medicine* 5, no. 7 (2008).

26. Marcel Hommel, "Towards a Research Agenda for Global Malaria Elimination," *Malaria Journal* 7, suppl. 1 (2008).

27. President's Malaria Initiative, "Malaria Operational Plan," November 13, 2009, http://www.fightingmalaria.gov/countries/mops/fy10/tanzania _mop-fy10.pdf.

28. Mark Grabowski, "The Billion Dollar Malaria Moment," *Nature*, February 27, 2008, http://www.nature.com/nature/journal/v451/n7182/full/ 4511051a.html.

29. "Program: Bill Gates III, *Keynote Address from Bill Gates III*, Co-Founder of Microsoft Corporation and the Bill & Melinda Gates Foundation," Come Together Washington, University of Washington Foundation, October 2004, http://uwfoundation.org/events_pubs/ctw_program_Bill GatesIII.asp.

CHAPTER 8

1. Sholto Byrnes, "The One to Watch in 2008," *New Statesman*, January 3, 2008, http://www.newstatesman.com/politics/2008/01/malaria-alonso -centre.

2. Ian Sample, "Malaria: GM Mosquitoes Offer New Hope for Millions," *The Guardian*, March 20, 2007; Mauro T. Marrelli, Chaoyang Li, Jason L. Rasgon, and Marcelo Jacobs-Lorena, "Transgenic Malaria-Resistant Mosquitoes Have a Fitness Advantage When Feeding on *Plasmodium*-Infected Blood," *Proceedings of the National Academy of Sciences*, March 19, 2007, http://www.pnas.org/content/104/13/5580.full.

3. Judith Epstein, "What Will a Partly Protective Malaria Vaccine Mean to Mothers in Africa?" *The Lancet* 370, no. 9598 (2007): 1523–1524.

4. Ibid.

5. Gooznews on Health, "The Dr. Ruth of Malaria Research," October 27, 2007, http://www.gooznews.com/archives/00844.html?page=300.

6. Russell had a career in the Army Medical Corps from 1959 through 1990, rising to commandant of the Walter Reed Army Institute of Research, and was founding president of the Sabin Vaccine Institute.

7. Nicholas A.V. Beare, Simon P. Harding, Terrie E. Taylor, Susan Lewallen, and Malcolm E. Molyneux, "Perfusion Abnormalities in Children with Cerebral Malaria and Malarial Retinopathy," *Journal of Infectious Diseases* 199, no. 2 (2009): 263–271, http://www.journals.uchicago.edu/ doi/abs/10.1086/595735?url_ver=Z39.88–2003&rfr_id=ori:rid:cross ref.org&rfr_dat=cr_pub%3dncbi.nlm.nih.gov.

8. Stanley Meisler, "Gaudi's Gift," Smithsonian.com, July 2002, http://www.smithsonianmag.com/arts-culture/gaudi.html.

CHAPTER 9

1. Bill Gates, "Remarks of Bill Gates, Harvard Commencement, 2007," *Harvard Gazette*, June 7, 2007, http://news.harvard.edu/gazette/story/2007/06/remarks-of-bill-gates-harvard-commencement-2007/.

2. James Surowiecki, "Push and Pull," *New Yorker*, December 20, 2004, http://www.newyorker.com/archive/2004/12/20/041220ta_talk_surowiecki.

3. Reuben Kyama and Donald G. McNeil, Jr., "Distribution of Nets Splits Malaria Fighters," *New York Times*, October 9, 2007, http://www.nytimes.com/2007/10/09/health/09nets.html.

4. Pedro Alonso, "Malaria: Deploying a Candidate Vaccine (RTS,S/ASO2A) for an Old Scourge of Humankind," *International Microbiology* 9 (2006): 83–93, http://www.im.microbios.org/0902/0902083.pdf.

5. UNICEF, "Mozambique: Statistics," March 2, 2010, http://www.unicef.org/infobycountry/mozambique_statistics.html; UN Office for the Coordination of Humanitarian Affairs, "'Lazarus Drug': ARVs in the Treatment Era," June 30, 2010, http://www.irinnews.org/InDepthMain.aspx?InDepthId=12&ReportId=56099.

6. Marc Herman, "Malaria: The View from Mozambique," GlobalPost, June 11, 2009, http://www.globalpost.com/dispatch/health/090317/malaria-mozambique?page=0,1.

CHAPTER 10

1. PATH Malaria Vaccine Initiative and Sanaria, "Phase I Trial of the Whole-Parasite Malaria Vaccine to Begin," press release, April 23, 2009, http://www.malariavaccine.org/files/04202009__SanariaP1trial_PR_FINAL.pdf.

2. Anjali Nayar, "Malaria Vaccine Enters Phase III Clinical Trials," *Nature News*, May 27, 2009, http://www.nature.com/news/2009/090527/full/news.2009.517.html.

3. Leslie Roberts, "Polio: No Cheap Way Out, *Science*, April 20, 2007, http://www.sciencemag.org/cgi/content/summary/316/5823/362.

4. Zoe Alsop, "Malaria Vaccine Trials Put Researchers to the Test," *Toronto Globe and Mail*, June 16, 2009, http://www.theglobeandmail.com/news/world/malaria-vaccine-trials-put-researchers-to-the-test/article1184851/.

5. Ibid.

6. Voice of America, "Malaria Parasite Becoming Resistant to Most Effective Medicine," March 5, 2009, http://www1.voanews.com/english/news/a-13-2009-03-05-voa24-68727807.html.

7. Simeon Bennett, "Malaria Disaster Looms from Bug's Resistance, Fakes," Bloomberg, May 26, 2009, http://www.bloomberg.com/apps/news?pid=conewsstory&tkr=NOTA:GR&sid=adA_6th.08Jk; Paul N. Newton, Rose McGready, Facundo Fernandez, Michael D. Green, Manuela Sunjio, Carinne Bruneton, Souly Phanouvong, et al., "Manslaughter by Fake Artesunate in Asia—Will Africa Be Next?" PLoS Medicine, June 13, 2006, http://www.plosmedicine.org/article/info:doi/10.1371/journal.pmed.0030197.

8. John Bingham, "Malaria May Be Developing Resistance to Main Drugs," The Telegraph, May 29, 2009, http://www.telegraph.co.uk/health/healthnews/5404735/Malaria-may-be-developing-resistance-to-main-drugs.html.

9. Bennett, "Malaria Disaster."

10. Judith Epstein, "What Will a Partly Protective Malaria Vaccine Mean to Mothers in Africa?" The Lancet 370, no. 9598 (2007): 1523–1524.

11. Government funding has come from the U.S. Department of Defense and the National Institutes of Health. In Sanaria's earliest days, funding also came from a band of friends and helpful lawyers, to whom Hoffman gave a small slice of equity in the company.

12. "Malaria 2010: More Ambition and Accountability Please," The Lancet 375, no. 9724 (2010): 1407, http://www.thelancet.com/journals/lancet/article/PIIS0140-6736(10)60601-0/fulltext#.

13. "Thomas Edison," Wikiquote, http://en.wikiquote.org/wiki/Thomas_Edison. As Wikiquote points out, there are many variations of this quotation.

CHAPTER 11

1. "America Philanthropy," The Economist, January 25, 2007.

2. "Bill Gates: World Economic Forum 2008," Remarks by Bill Gates, Chairman, Microsoft Corporation, "A New Approach to Capitalism in the 21st Century," World Economic Forum 2008, Davos, Switzerland, January 24, 2008, text online at Microsoft News Center, http://www.microsoft.com/presspass/exec/billg/speeches/2008/01-24wefdavos.mspx.

3. Jeffrey Bradach, "Going to Scale," Stanford Social Innovation Review, Spring 2003, Bridgespan Group, http://www.bridgespan.org/LearningCenter/ResourceDetail.aspx?id=484&Resource=Articles.

4. Patrick Adams, "The Sanaria PfSPZ Malaria Vaccine, Until Recently Considered Impossible, Is Entering Phase II Trials," TropIKA.net, December 16, 2009, http://www.tropika.net/svc/interview/Adams-20091216-Interview-Hoffman.

5. Bloomberg Businessweek, "2009 Best Places to Launch a Career," September 3, 2009, http://images.businessweek.com/ss/09/09/0903_places _to_launch_a_career/8.htm.

6. Jonathan Kozol, *On Being a Teacher* (Oxford, U.K.: OneWorld Publications, 1981).

7. Melinda French Gates, "Malaria Forum Keynote Address," transcript at Bill & Melinda Gates Foundation website, October 17, 2007, http://www .gatesfoundation.org/speeches-commentary/Pages/melinda-french-gates -2007-malaria-forum.aspx.

CHAPTER 12

1. "Counting Malaria Out," Statement by Professor Awa Marie Coll-Seck, Executive Director, Roll Back Malaria Partnership, Delivered at Opening of Photo Exhibition "Malaria: Blood, Sweat and Tears," United Nations Headquarters, April 22, 2010, http://www.rbm.who.int/globaladvocacy/ st2010-04-22.html; Ray Chambers, "The Numbers Tell a Promising Story for World Malaria Day 2010," *Financial Times*, April 22, 2010, http://www.ft .com/cms/s/0/a135a164-4ce6-11df-9977-00144feab49a.html.

2. World Health Organization, "Summary," World Malaria Report 2009, http://whqlibdoc.who.int/publications/2009/9789241563901_eng.pdf, p. viii.

3. P. J. Hotez and A. Kamath, "Neglected Tropical Diseases in Sub-Saharan Africa: Review of Their Prevalence, Distribution, and Disease Burden," *PLoS Neglected Tropical Diseases* 3, no. 8 (2009): e412, doi:10.1371/ journal.pntd.0000412.

4. World Health Organization, "One Page Fact Sheet for World Malaria Report 2009," http://www.who.int/malaria/world_malaria_report_2009/ factsheet/en/index.html.

5. "Malaria 2010: More Ambition and Accountability Please," *The Lancet* 375, no. 9724, (2010): 1407, http://image.thelancet.com/journals/ lancet/article/PIIS0140-6736(10)60601-0/fulltext.

6. Robert A. Guth, "Gates Rethinks His War on Polio," *Wall Street Journal*, April 23, 2010, http://online.wsj.com/article/SB100014240527023033 48504575184093239615022.html.

7. Ibid.

8. Ibid.

9. S. L. Hoffman, P. F. Billingsley, E. James, A. Richman, M. Loyevsky, T. Li, S. Chakravarty, et al., "Development of a Metabolically Active, Non-Replicating Sporozoite Vaccine to Prevent Plasmdium Falciparum Malaria," *Human Vaccines* 6, no. 1 (2010): 97–106, http://www.ncbi.nlm.nih.gov/ pubmed/19946222.

10. "Phase I Trial of Whole-Parasite Malaria Parasite to Begin," press release, Sanaria and PATH Malaria Vaccine Initiative, April 23, 2009.

11. Richard Neustadt, *Presidential Power* (New York: John Wiley and Sons, 1960); Charles O. Jones, "Professional Reputation and the Neustadt Formulation," *Presidential Studies Quarterly*, June 1, 2001, http://www.access mylibrary.com/coms2/summary_0286-10431316_ITM.

12. Warren Bennis, "Only the Optimists Survive," Bloomberg Businessweek, http://www.businessweek.com/managing/content/may2009/ca20090 518_917239.htm.

13. Fifteen of the pictures from this event, by photographer Adam Nadel, can be seen at "Week in Review: Where Malaria Still Kills," *New York Times* website, http://www.nytimes.com/slideshow/2010/02/21/weekinreview/ 20100221-malaria-ss_2.html.

14. Ibid., photo 2 of 15.

15. "Ophelia Dahl's Commencement Address to the Wellesley College Class of 2006," http://www.wellesley.edu/PublicAffairs/Commencement/ 2006/ODahl.html.

INDEX

Bill Shore is the founder and executive director of Share Our Strength®, the nation's leading organization working to end childhood hunger in America, which he founded in 1984. He currently serves on the board of  directors of The Timberland Company, and of Venture Philanthropy Partners. Shore has been an adjunct professor at New York University's Stern School of Business and the program advisor for the Reynolds Foundation Fellowship program at the John F. Kennedy School of Government's Center for Public Leadership. He is the author of three books.

PublicAffairs is a publishing house founded in 1997. It is a tribute to the standards, values, and flair of three persons who have served as mentors to countless reporters, writers, editors, and book people of all kinds, including me.

I. F. STONE, proprietor of *I. F. Stone's Weekly*, combined a commitment to the First Amendment with entrepreneurial zeal and reporting skill and became one of the great independent journalists in American history. At the age of eighty, Izzy published *The Trial of Socrates*, which was a national bestseller. He wrote the book after he taught himself ancient Greek.

BENJAMIN C. BRADLEE was for nearly thirty years the charismatic editorial leader of *The Washington Post*. It was Ben who gave the *Post* the range and courage to pursue such historic issues as Watergate. He supported his reporters with a tenacity that made them fearless and it is no accident that so many became authors of influential, best-selling books.

ROBERT L. BERNSTEIN, the chief executive of Random House for more than a quarter century, guided one of the nation's premier publishing houses. Bob was personally responsible for many books of political dissent and argument that challenged tyranny around the globe. He is also the founder and longtime chair of Human Rights Watch, one of the most respected human rights organizations in the world.

· · ·

For fifty years, the banner of Public Affairs Press was carried by its owner Morris B. Schnapper, who published Gandhi, Nasser, Toynbee, Truman, and about 1,500 other authors. In 1983, Schnapper was described by *The Washington Post* as "a redoubtable gadfly." His legacy will endure in the books to come.

Peter Osnos, *Founder and Editor-at-Large*